COMMUNICATING
SCIENCE
TO THE PUBLIC

The Ciba Foundation is an international scientific and educational charity. It was established in 1947 by the Swiss chemical and pharmaceutical company of CIBA Limited—now CIBA-GEIGY Limited. The Foundation operates independently in London under English trust law.

The Ciba Foundation exists to promote international cooperation in biological, medical and chemical research. It organizes about eight international multidisciplinary symposia each year on topics that seem ready for discussion by a small group of research workers. The papers and discussions are published in the Ciba Foundation symposium series. The Foundation also holds many shorter meetings (not published), organized by the Foundation itself or by outside scientific organizations. The staff always welcome suggestions for future meetings.

The Foundation's house at 41 Portland Place, London, W1N 4BN, provides facilities for meetings of all kinds. Its Media Resource Service supplies information to journalists on all scientific and technological topics. The library, open seven days a week to any graduate in science or medicine, also provides information on scientific meetings throughout the world and answers general enquiries on biomedical and chemical subjects. Scientists from any part of the world may stay in the house during working visits to London.

Ciba Foundation Conference

COMMUNICATING SCIENCE TO THE PUBLIC

A Wiley – Interscience Publication

1987

JOHN WILEY & SONS

Chichester . New York . Brisbane . Toronto . Singapore

Published in 1987 by John Wiley & Sons Ltd, Baffins Lane, Chichester, Sussex PO19 1UD, UK.

Ciba Foundation Conference
× + 214 pages, 18 figures, 13 tables

Library of Congress Cataloging-in-Publication Data

Communicating science to the public.

'Editors: David Evered (organizer) and Maeve O'Connor'—Contents.
Papers presented at the Conference on the Communication of Science, held at the Ciba Foundation, London, 14–16 Oct. 1986.
'A Wiley–Interscience publication.'
Includes index.
1. Communication in science—Congresses.
2. Science news—Congresses. I. Evered, David.
II. O'Connor, Maeve. III. Conference on the Communication of Science (1986 : Ciba Foundation)
Q223.C6544 1987   501'.4   87–8261

ISBN 0 471 91511 4

British Library Cataloguing in Publication Data

Communicating science to the public.—
(Ciba Foundation conference).
1. Science news
I. Ciba Foundation II. Series
507   Q225

ISBN 0 471 91511 4

Typeset by Inforum Ltd, Portsmouth
Printed and bound in Great Britain.

Contents

Participants

B. Bell   Curriculum Development Division, Department of Education Private Bag, Wellington, New Zealand

Sir Walter Bodmer   Imperial Cancer Research Fund Laboratories, PO Box 123, Lincoln's Inn Fields, London, WC2A 3PX, UK

S. Caravita   Istituto di Psicologia del CNR, Reparto: Psicopedagogio, Via U Aldrovandi 18, 00197 Roma, Italy

G. Deehan   Room 7079, BBC Radio, Broadcasting House, Langham Place, London, W1A 1AA, UK

G. Delacôte   CNRS, 15 Quai Anatole France, 75700 Paris, France

B. Dixon   7 Warburton Court, Victoria Road, Ruislip Manor, Middlesex, HA4 0AN, UK

S. Duensing   The Exploratorium, 3601 Lyon Street, San Francisco, California 94123, USA

D.C. Evered   The Ciba Foundation, 41 Portland Place, London, W1N 4BN, UK

J.H. Falk   NHB MRC 101, Smithsonian Institution, Washington, DC 20560, USA

A.J. Friedman   New York Hall of Science, 47-01 111th Street, Corona, New York 11368, USA

R.L. Gregory   Brain & Perception Laboratory, Department of Anatomy, University of Bristol, The Medical School, University Walk, Bristol, BS8 1TD, UK

J.P. Hearn   Institute of Zoology, Regent's Park, London, NW1 4RY, UK

W.M. Laetsch   University of California, Berkeley, California 94720, USA

A.M. Lucas   Centre for Educational Studies, Kings College London (KQC), Chelsea Campus, 552 King's Road, London, SW10 0UA, UK

M. Macdonald-Ross   Institute of Educational Technology, Open University, Walton Hall, Milton Keynes, MK7 6AA, UK

R.S. Miles   Department of Public Services, British Museum (Natural History), Cromwell Road, London, SW7 5BD, UK

J.D. Miller   Public Opinion Laboratory, Northern Illinois University, DeKalb, Illinois 60115-2854, USA

D. Nelkin   The Program for Science, Technology & Society, 632 Clark Hall, Cornell University, Ithaca, NY 14853, USA

S. Quinn   The Science Museum of Minnesota, 30 East 10th Street, Saint Paul, Minnesota 55101, USA

M. Rhodes   BBC TV, Kensington House, Richmond Way, London, WI4 0AX, UK

C.G. Screven   Department of Psychology, University of Wisconsin-Milwaukee, PO Box 413, Milwaukee, Wisconsin 53201, USA

B. Serrell   Serrell & Associates, 5429 South Dorchester, Chicago, Illinois 60615, USA

I.E.F. Stewart   CIBA-GEIGY plc, 30 Buckingham Gate, London, SW1E 6LH, UK

L. Tan Wee Hin   Singapore Science Centre, Science Centre Road, Singapore 2260

G.P. Thomas   Department for External Studies, University of Oxford, Rewley House, 1 Wellington Square, Oxford, OX1 1JA, UK

C.A. Tisdell   Department of Economics, University of Newcastle, New South Wales, 2308 Australia

R.D. Whitley   Faculty of Business Administration, Manchester Business School, Booth Street West, Manchester, M15 6PB, UK

L. Wolpert   Department of Anatomy & Biology as Applied to Medicine, The Middlesex Hospital Medical School, Cleveland Street, London, W1P 6DB, UK

A basis for better public understanding of science

W.M. LAETSCH

Chancellor's Office, University of California, Berkeley, CA 94720, USA

Abstract. Pleas for improved formal and informal science education are urgent and widespread. These pleas assume that communicating the facts and principles of science to the general public is critical to societal welfare. The main claims are that (1) a knowledge of science will permit the electorate to make better political decisions; (2) understanding the basis of modern technology brings economic returns and promotes national security; (3) scientific knowledge will eliminate superstition and non-rational views of the universe; (4) behaviour will improve if consequences are known; (5) familiarity with the scientific method will lead to a more ethical world view. The reasoning supporting these claims, however, is flawed and does not provide a proper basis for the effective long-term communication of science. The future health of science communication rests instead on understanding that science is interrelated with other elements of our culture, on understanding that curiosity about the physical and biological world is a fundamental trait whose satisfaction is a legitimate end, and on the realization that understanding science leads to an enriched existence. Science literacy is primarily a humanistic endeavour and should be communicated on that basis.

1987 Communicating science to the public. Wiley, Chichester (Ciba Foundation Conference) p 1–18

Science education is a high priority for most nations. It is self-evident that it is essential to providing the scientists and technicians required for societies which are, or wish to become, industrialized. Rapid changes in industrial technology, weaponry, health care, agriculture, communications and the environment have highlighted the need for a large and steady stream of highly trained scientists and engineers. This conference, however, is really concerned with the education of those who are not or will not be professional scientists and engineers and highly trained technicians.

Historical analysis of the rationale for efforts to communicate science to the non-scientist is a topic in itself. Those efforts are of long standing, as testified by the activities of schools, museums, zoos, and the printed and electronic media represented at this conference. Although earlier motives for communicating science to the public have not been carefully compared with current

1

motives, to my knowledge, many long-established museums, zoos, and bota-
nical gardens were originally charged with exhibiting objects and transmitting
knowledge for general edification and to satisfy the curiosity of the general
public. The current goal is to promote scientific literacy. This commonly
expressed but poorly defined state of being gained currency in the USA some
time after the launching of Sputnik in 1957. This event sparked the rapid
infusion of federal funds into the development of pre-college science curricula. While the felt need to regain the lead in space technology provided the
political muscle for the expanded movement towards curriculum reform, the
result was a significant investment in communicating science to young people.
A good argument can be made that these programmes—particularly those
involved with science in elementary schools—provided intellectual orientation for many scientific literacy programmes for the general public. Because
of the requirements of funding sources, the need to justify science for the
general good in political terms has always been present. This has been
particularly true whenever governmental largess has diminished.

The entwining of science and public policy issues added another dimension
to science education. Nuclear power, the energy crisis, the environmental
movement and preventive health care have all spawned educational programmes which are usually funded by the government. It is hardly surprising that
the rationale for science education is defined by the relation with political and
economic issues. The case for school reform in the USA, at least since
Sputnik, has been phrased in the utilitarian terms that someone is getting
ahead of us in the military or commercial sector. *A Nation at Risk* (National
Commission on Excellence in Education 1983), one of the best known of the
recent calls in the USA for school reform, even used the terms of the arms
race—'unilateral educational disarmament'—to describe our real or imagined
relationship to commercial competitors.

Definitions of scientific literacy

Scientific literacy is a valuable commodity for educators, but what is it? An
entire issue of *Daedalus* in 1983 struggled with definitions. Most definitions
lean heavily on experiencing processes of science rather than on learning
about a specified body of knowledge. The list compiled by Harrison (1982) is
fairly typical.

The expectations we have for the scientifically literate are often impressive.
Our desire seems to be for amateur philosopher-kings who possess wisdom
and an understanding of the relationship between science and other domains
beyond that of the professional scientist. For example, anyone who reviews
manuscripts presented for publication by experienced scientists is familiar
with the widespread confusion between causality and correlation and with the
problem of drawing inferences from observations.

The claims for scientific literacy

While definitions of scientific literacy vary, I am most interested here in the claims made for what it achieves. Every report and editorial on the nature and need for better science education and every proposal to a granting agency to improve science education that I have ever seen has presented similar claims. The following references are typical. The weakness in science and mathematics education threatens the country's economic and military security (Reagan 1982). 'Improved preparation of all citizens in the fields of mathematics, science, and technology is essential to the development and maintenance of our Nation's economic strength, military security, commitment to the democratic ideal of an informed and participating citizenry, and leadership in mathematics, science, and technology' (National Science Board 1982). 'The deficiency in the quality and quantity of teaching of science and mathematics—subjects that are emphasized in a number of countries that are our competitors—is undoubtedly a factor in our country's economic decline. Lack of scientific literacy threatens the efficient, or even adequate, function of our democracy in this scientific age' (Seaborg 1983).

The primary reasons given for promoting scientific literacy are (in approximate priority order): (1) a knowledge of science will result in the electorate making better political decisions; (2) understanding the basis of modern science and technology brings economic returns; (3) scientific knowledge will eliminate superstition and non-rational views of the universe; (4) scientific knowledge defines consequences of behaviour and knowledge will change behaviour; (5) familiarity with the scientific method will lead to a more ethical world view.

Implicit in these and similar claims is a high level of hubris. It is assumed that scientific literacy is the highest literacy and those who are deficient are less useful members of society.

Do these near-ubiquitous claims, which are presented both in the popular press and in publications devoted to scientific matters, evolve from the application of scientific methodology? I submit that they will not bear scrutiny and are seriously flawed. As such, they cannot be the basis for supporting a long-term commitment to communicating science. Each of these claims will be examined.

(1) Better political decisions?

The pre-eminent claim is that we must have a scientifically literate citizenry because most public policy issues have a science or technology component. Informed voters will make better political choices. This argument states that the absence of widespread scientific literacy will leave the debates to a scientific priesthood, and that subsequent political decisions will be driven by an elite few whose actions will erode democratic institutions.

This pre-eminent claim is also the easiest to topple. If true, then those most literate about science, the professional scientists, should provide evidence of possessing superior political wisdom. There is no evidence that professional scientists vote in a more enlightened manner than do reasonably intelligent people who are illiterate in science. Any serious probing of this dichotomy would probably reveal that the advantage goes to the latter group.

This claim contains the basic fallacy that scientific facts are absolute and understanding the facts will lead to uniform interpretations. If the debates on interpretations of fact that provide the substance of meetings of scientists are not enough to demonstrate the fallacy, the public debates over nuclear energy, toxic wastes, cancer-causing agents and the relationship between nutrition and health should do so. In each case, batteries of professional scientists hammer each other over the interpretation of frequently agreed-upon facts. Legislators frequently explode with frustration over the experts who cannot agree on the meaning of the same information. Voters frequently respond by sitting on their hands in deference to their instincts to distrust squabbling experts—a wise practice unlikely to be enhanced by increased degrees of scientific literacy.

Underlying this claim is the assumption that scientific knowledge and not politics drives legislative decisions. An article of faith possessed in inordinate degrees by physical scientists in particular is that all problems have solutions, that definition of the problem will lead to a solution, and that presentation of the logic of a position is sufficient to persuade. That this is contrary to all empirical evidence of the legislative process is lost on those who make claims that scientific literacy will lead to political literacy.

Another problem about claiming that scientific literacy promotes better political decisions revolves around the definition of 'better'. The ethical, economic, legal and religious definitions of what is better are determinants in political decisions. Outside well defined health issues, scientific logic is rarely the determining factor in defining what is better. Even in the development of weapons, the best is more a function of interservice rivalries, regional politics, loyalties of senior politicians, and the influence of contractors than it is of scientific and technical wisdom. Universal scientific literacy will not change the equation.

The claim that scientific literacy results in more efficacious political behaviour reflects the belief that knowledge and progress are positively correlated. There are many examples, however, in which real or imagined scientific knowledge turns on itself. A curious example is the fact that a number of leaders of the creationist movement in the United States have Ph.D. degrees in scientific disciplines and are certainly literate about science. Many current disputes over products of genetic engineering are further testaments. A case study consists of the so-called ice-free bacteria developed at Berkeley. Some bacteria living on the surface of leaves produce a surface protein supposedly providing the nucleus for the formation of ice crystals which damage the plant. Gene cloning has produced a bacterial strain without this protein and ice does not form. Field trials in which the new strain would be applied to plants have been blocked by political activists backed by scientists who claim that the introduction of genetically altered organisms into the environment presents unacceptable hazards. Many would argue that this is a prime example of some scientific knowledge being disfunctional. It might also be said that opposition to such testing has a science-oriented veneer, masking resentment to being excluded from the decision-making process. This may be a function of public relations illiteracy on the part of scientists rather than lack of real scientific literacy on the part of the public.

The use or misuse of some ecological principles in the development of the environmental movement is a further example of the impact of knowing some science. It has resulted in the common current belief in the balance of nature which is at the root of many fears of using genetically altered organisms produced by genetic engineering.

The field of nutrition provides a battleground for experts (Marshall 1986) and also provides innumerable examples of how some understanding of science has led to general confusion. The problem is illustrated by the many people who believe they are scientific literates but also believe that margarine made of unsaturated fats has fewer calories than butter, or that nuts have fewer calories and are better for you than an equivalent weight of pizza.

Even if it were true that scientific literacy leads to informed decision making, at what point in the literacy spectrum does this happen? At what point does information cease to promote misunderstanding?

(2) Better economic returns?

Assumptions about the positive relationship between scientific literacy and the political process are most commonly espoused by educators and scientists. Politicians, on the other hand, espouse the economic relationship—a relationship at the heart of the allocation of funds. The governor and legislature of the State of California make no bones about their support of the massive research enterprise of the University of California, because of the evidence that university research provides the basis for the economic wellbeing of the state. The same applies at the national level, and support for formal science education is founded on the necessity for a proper supply of scientists and engineers. It is not really clear, however, whether there is a causal relationship between the health of formal science education and the supply of professionals. The state of school science education in the US has been considered to be in a crisis since Sputnik—and by implication for a long period before that—but the number of young people wishing to be engineers, doctors and scientists has fluctuated greatly during this period. Employment opportunities are a much greater determinant of supply than is excellence of science instruction. Many young people, and their parents, now equate both science and engineering with computers, and the relatively high salaries in the computer field, combined with media advertising and programming, have made becoming something called a computer scientist the goal for millions of young people. We cannot accommodate the demand for university courses and if we could, the marketplace would be saturated. The argument here is that bright students aggregate according to opportunities, whether they be economic or intellectual.

Does scientific literacy for the general public generate economic returns? I doubt it. People who manufacture computers do not have to be scientifically literate. Certainly the balance of payments problem in the USA and the ability of British industry to compete has little to do with the scientific literacy of our labour force or with that of our competitors. Several years ago there was debate in Britain about scientific literacy and economic health. The literacy was considered to be high but the economy was in the doldrums. Apparently another kind of literacy or cultural ethic was required. Whether the new version could help the exchange rate of the pound is undetermined.

(3) Less superstition?

A fond hope of our rational world view is that scientific knowledge will eliminate superstition and what we consider non-rational interpretations of the universe. Biologists rail against creationists, astronomers against astrologers, and health scientists against quacks. All of them say that if only people knew enough of the right things, they would see the error of their ways. It is a vain hope and one which defies vast amounts of empirical data accumulated over a long period. It ignores what can be observed on a daily basis—that people can simultaneously hold different belief systems. The various systems can run their separate courses, or one or the other can predominate. A scientist can spend six days of each week exploring facets of the rationally interpreted universe and extolling the immutable laws of that universe, and on the seventh day pray to an anthropomorphic deity to overturn those very laws. Educated people throughout the world use and believe in both western and traditional systems of medicine. The spectacular advances of western medicine have not had an appreciable effect—at least in the United States—on the practice of faith-healing. The scientifically literate in some cultures would not consider letting their children marry without consulting astrologers. The list of such paradoxes is endless. A failure to recognize these eternal dualities as a fact of the human condition is due to cultural illiteracy and the hubris of science. An excellent discussion of the relationship of non-rational beliefs to science is given by Singer & Benassi (1981) and Alan Friedman discusses this topic in this conference.

(4) Improved behaviour?

Another reason for obtaining scientific literacy is that it will reinforce positive behaviour as defined by the behaviour's consequences. This certainly works for a multitude of cause-and-effect relations encountered daily. The germ theory of disease has been a great persuader and continues to modify behaviour, but our knowledge of many other consequences related to human health—for example, the effects of diet, alcohol and tobacco (Kolata 1986)—has had little effect on behaviour. The uses of water, energy resources and land provide myriad examples of the failure of knowledge to determine behaviour. It would be trite to make this argument except that claims for behaviour alteration are constantly made in support of the need to communicate science.

(5) A more ethical world view?

Does scientific literacy promote a more ethical and humane world view, as a typical claim suggests (Hamburg 1986)? The practice of science has shared assumptions because science is done in a prescribed fashion. What might be called a shared ethic is maintained by the accountability resulting from duplication of experiments. This actual or threatened verification of evidence is an immensely powerful force which gives the impression that science is honest, and while individual scientists are no more honest than anyone else, the practice of science is remarkably honest. Like the work of airport traffic controllers, the work of scientists can be verified. The assumption, however, that knowledge of science will lead to a more ethical world view is fallacious. Even a cursory glance at the behaviour of nations with high levels of general and scientific literacy is enough to dispel this notion.

The benefits of scientific literacy

Americans have strong reformist tendencies, so they love crises, and every decade or so we roll out an educational crisis whose solution is essential to keep the Russians or Japanese at bay. But over the long run we cannot continue to communicate science on the basis that it will maintain both our democracy and our technological edge. The rationale won't stand up to scrutiny and in the meantime we offer false claims to the granting agencies and hope they play the crisis game.

What can we claim are the benefits for being scientifically literate? Before answering, I will return to a previous theme on the hubris of scientific literacy, i.e. that this is the preferred literacy and that it must be obtained to get a job, understand our technological society, and understand the world around us. This is rarely challenged, in spite of the fact that vast numbers of people lead useful and happy lives and are very uninformed in any formal sense about science. Why is it better to be literate about science than about the arts, politics, ethics, or even public relations? Graubard (1983) illustrates this issue and balances the scientific literacy cause with a historian's insight. He also punctures some of the arguments about the superiority of educational systems of competing nations.

Generic literacy, not necessarily a specific form, is the goal. People can and will follow the most congenial muse. Given the opportunity, much of what we call scientific literacy can take care of itself. It probably does so, if in somewhat different form than usually presented. For example, there are supposed to be about 30 million rose-growers in the United States, and a substantial number of these are serious about their avocation. You cannot be a serious grower of roses—or other plants for that matter—without knowing a fair amount about practical science. The same can be said for the dozens of other hobbies with a science orientation. Yet we never hear that the reason for growing roses is to better understand public policy issues.

Rose-growers provide leads to the reasons for people's interest in science, and hence to the basis for its communication. No one can stop the interest in science. The millions of amateur astronomers, bird-watchers, gardeners, rock collectors, and electronic gadgeteers are only a portion of those avidly spending time at science. People demand science museums, zoos and nature reserves. Everyone who fishes or hunts is a student of animal behaviour. Those people's interest in the world around them is inexorable. This is seen most clearly in children. Their curiosity about the natural world and how things work is insatiable. During the formal process of imposing scientific literacy on them, we often force all but the pre-professionals to hide this curiosity for a period. Once they are released from formal requirements, their curiosity leads them forth once again to explore science. Curiosity is so obviously a fundamental and powerful characteristic of our species that we fail to take it seriously. People go to science museums and science centres and zoos and watch science programmes on television to satisfy their curiosity. This is the primal motive. They are not generally interested in science as a formal discipline. It is not an imperative for them but is only one element in the fabric of their culture.

Fulfilling curiosity about objects and activities is considered an end in itself for visitors to institutions concerned with informal science education. These institutions know the value of stimulating and fulfilling curiosity by means of active exploration and they increasingly focus their efforts to these ends. They know this is a justifiable end but they hesitate to admit this publicly—or to granting agencies. The traditional hair shirt of education still partially covers their intentions. Whether exhibits or programmes are frivolous or formally academic, it is too often stated that they will promote 'problem-solving' or prepare people to participate fully in our technologically based society. The stated goals of many science museums are strictly utilitarian (Carnes 1986).

What a delight it would be to send a proposal for funding to a granting agency for an exhibit that promised to stimulate curiosity and that would be judged on whether it did so. Increased scientific literacy would be an inevitable by-product. It is time to take a message from cultural institutions concerned with art and history. They are what they are and they prosper because or in spite of that. Viewing art or listening to music is considered valuable in its own right. These activities are not a means to an external end in the way that scientific literacy is often portrayed as a means to economic splendour.

There is a very important reason for promoting scientific literacy. It makes our lives far richer than is possible without a knowledge of the surrounding world. We are literally blind and deaf and without a sense of smell if we are not able to understand common natural events, to know other living things, and to explain what we see and to make some predictions based on what we know. As an example, I am a botanist and when I am in the fields and forests of much of North America, I am reading a familiar story. I know what is related and what is not and what might have happened yesterday and what will happen today, tomorrow and next month. This is intensely satisfying in what is essentially an aesthetic sense. When I am in an unfamiliar tropical forest, the environment is much less rich to me even though it is biologically much richer. I don't know the names of most of the organisms or how they work. There is a sense of wonder, but I am not as enriched. This same desire for enriching the environment by knowing is observed in zoos and museums. Knowing the name creates the animal or object. The ability to observe and understand the world around us makes our lives better because we fulfil our humanness. Understanding science for its own sake is ample and sufficient reason for promoting scientific literacy.

This might be regarded as a quaint reaffirmation of 19th century reasons for pursuing science. I might be further accused of pursuing a romantic ideal reflected by a 19th century leisured class devoting themselves to amateur science. This objection has a point, but not the one intended. People in the western world have enormous amounts of leisure. They retire from work early and they remain active for a larger portion of their lifespan. Seen in this light, the disinterested pursuits of the Victorians are very relevant to current times and to the pursuit of scientific literacy.

Science as a body of knowledge is not different for professional scientists and members of the public, so reasons for being interested in science have no need to differ. Even though scientists like to tell granting agencies that their work will help the economy or save lives, they do science because it is intellectually exciting. They are curious and want to know more because it is satisfying—often in a spiritual sense.

Scientific literacy helps us to understand ourselves and our environment and the relationship between the two. The ultimate rationale for scientific literacy is humanistic. We should present this without apology and be consistent in our advocacy. Over time the support for science education will be more constant and science will be more respected.

The reluctance to espouse scientific literacy as a humanistic enterprise stems in large part from literacy gaps in other areas of human endeavour. Is it possible to really understand science without understanding some philosophy, history, and even sociology? Science must be looked at apart before its role in our culture can be appreciated. Cultural literacy is required for scientific literacy to have much meaning. The latter is a symbiotic element and grows and functions best within, rather than outside, its cultural relationship.

If scientific literacy is perceived as one element of culture, one which promotes personal pleasure and growth rather than national security, then there are interesting implications for methods of science communication as well as for what is communicated. It should be easier to make the communication of science effective if we agree on why we are doing it in the first place.

References

Carnes A 1986 Showplace, playground or forum? Choice point for science museums. Museum News 64:29–35
Daedalus 1983 Scientific literacy 112:1–251
Friedman A 1987 The influence of pseudoscience, parascience and science fiction. This volume, p 190–196
Graubard SR 1983 Nothing to fear, much to do. Daedalus 112:231–248
Hamburg DA 1986 New risks of prejudice, ethnocentrism, and violence. Science 231:533
Harrison AJ 1982 Goals of science education. Science 217:109
Kolata G 1986 Reducing risk: a change of heart? Science 231:669–670
Marshall E 1986 Diet advice, with a grain of Salt and a large helping of Pepper. Science 231:537–539
National Commission on Excellence in Education 1983 A nation at risk. United States Department of Education, Washington, DC
National Science Board 1982 Today's problems, tomorrow's crises. A report of the National Science Board Commission on Pre-college Education in Mathematics, Science, and Technology, National Science Foundation, Washington DC
Reagan R 1982 In an age of technology, the three R's are not enough. New York Times, May 16 1982
Seaborg GT 1983 A call for educational reform. Science 221:219
Singer B, Benassi VA 1981 Occult beliefs. American Scientist 69:49–55

DISCUSSION

Tisdell: A lot of communication in science is intended to obtain political support for the funding of science. From time to time universities ask their scientists to give them a few hundred words on this and that, so that they can distribute these and curry favour with the electorate.

Another reason for communicating science is to obtain public support for technology and scientific research that might otherwise not be socially acceptable. It is important to convince the public that nuclear energy, new fertilization techniques and so on are not as risky as they may think. The communicators may even bias the information they supply, making it into propaganda, or disinformation.

Industry is another important communicator of science and technology. Companies need to call the attention of their customers or sometimes their suppliers to new technology and new ideas.

Why do people communicate and why ought they to communicate? It may all be a matter of culture, as Dr Laetsch says, but an economist might say that if this information is for your personal benefit why shouldn't you pay for it? Why shouldn't you pay to go to a museum, for example? Maybe the government ought to pay but that would be another issue.

Falk: Are you concerned that scientific literacy is being put forth as a form of propaganda, Dr Laetsch, or that those who are issuing the propaganda actually believe it?

Laetsch: I am concerned about both of those aspects.

Dixon: I agree that science enriches our appreciation of the world but I am disturbed that you seem to think that, because that is how things are in the world, we must accept that public issues are not usually debated or decided on rational grounds. Recently there was concern in Europe, mostly mobilized by the European Consumers Association, about dangers that might arise from the use of steroid hormones to fatten animals more quickly. The committee set up to look into this decided there were no dangers at all. Their advice was in line with advice and recommendations of the Food and Drug Administration in the USA and also with the recommendations of the World Health Organization. But a few months ago the European agricultural ministers decided to ban the hormones. One of the agricultural commissioners is on record as saying that more attention has to be paid to political realities than to the facts of science. By political realities he meant what most people think. In this case what most people think is not based on any rational, objective or scientific grounds but on prejudice or irrationality. How does what you were saying apply to that case? Do you welcome that as being the way things are?

Laetsch: I should have emphasized the value of free discussion on this kind of matter. Certainly political decisions are made contrary to scientific and technical advice. That is unfortunate but it is to be expected and we should not be surprised. A scientific answer will not necessarily lead to a particular political decision. Even if people are provided with all the necessary information it may not determine how they vote.

The problem with the frost-free bacteria experiments I mentioned was not really an objection to the science but to the fact that the bioengineering company behind this did not bring local people into the political process. The local people then used the idea that the science would be harmful as an excuse to object to the project. The company involved didn't communicate effectively what they wanted to do.

Dixon: The frost-free bacteria pose a very complicated problem for someone who has no scientific understanding at all. At a meeting of the American Society for Microbiology in 1985 ecologists, molecular biologists and others spent a week thrashing out what might happen if this organism was released. It worries me that there is no short-term method of getting over that sort of background information easily.

Bodmer: Everybody here would support your humanistic arguments, Dr Laetsch, but I was taken aback by your attitude to the promotion of ignorance. You seemed to be setting up a straw man and saying that if people know a bit more about science and understand it a bit better, all decisions will necessarily be clearer. That is not what anybody says. You are saying that it might contribute to some improvement in the decision-making process, but where is the evidence? Anecdotes do not help us to look scientifically at the evidence, if there is any, that a better understanding of some aspects of science might contribute to better decisions. Your rather casual dismissal of many of the arguments is quite disturbing.

Laetsch: I am all in favour of scientific literacy. Knowing more science is better than knowing less. I was trying to state some basic kinds of values and reasons which I find are rarely stated.

Bodmer: We should be talking about evidence, not beliefs, especially when we are talking about the public understanding of science.

Laetsch: That is interesting in itself. One of the aspects of public understanding is to get across the idea that there are beliefs in science. Scientists who are experts and who are on opposite sides of any particular policy issue can use the same facts to come to quite different interpretations. Politicians then rub their hands and say 'a plague on both your houses'.

Bodmer: Public policy issues and the assessment of scientific results are two quite different things. Scientists will argue with each other over scientific results involving their intuition and aspects of belief but there is an enormous difference between taking the accepted consensus in a particular area and then saying what implications that has for a particular public policy issue.

Nelkin: The credibility of the scientific community is an important aspect of the process of communication. I have recently been studying the AIDS dispute and the communication of scientific information about transmissibility. The way that information is communicated bears a great deal on its credibility. If people do not feel part of a decision, scientists are not believed and we reach the dilemma you are talking about. The AIDS discussion, especially in schools, is a good example of an issue which ends up being about political rights.

Bodmer: Are you arguing that the information would not be better received if people understood a little more?

Nelkin: Communities where people were appropriately informed and felt involved in the decision to allow children with AIDS to attend school accepted that decision. Communities where the decision was announced as a *fait accompli* completely rejected the decision even though the scientific consensus was very strong that AIDS is not transmissible except through body fluids. A key variable is the way scientific information, which is the same in every case, is conveyed to the public.

Bodmer: That already says that it was better to convey the information than not to. But conveying it inadequately did not help.

Wolpert: You are saying that being scientifically literate does not solve all the problems, therefore why bother to be scientifically literate. Dorothy Nelkin's point is the same—that being scientifically literate often doesn't resolve how you are going to treat your children when someone with AIDS goes to their school. But there may still be many situations where being scientifically literate will help. If everyone was scientifically illiterate we would be culturally diminished; we would also be hampered in the way we look at the world and make judgements about what is going on.

We must distinguish between the process of science and scientific knowledge. It is important that people should have not only some knowledge of science but also some feel about the process of science. Without that they cannot really judge what scientists tell them. A lot of what goes on in our society is science-based. I am very interested in the anti-science movement and one small way of countering the anti-science movement is for people to understand how science works. There are all sorts of subtle and important areas in which being scientifically literate makes you a better person, helps you to make better political judgements, and makes you behave in a better ethical way, although I have no scientific evidence for this. But your position that if scientific literacy doesn't do everything it is useless is very dangerous.

Whitley: We have already come to a demarcation dispute—what is a scientist? We need some historical background here. We are all talking about science as if we know what it is and as if it is the same today as it was when the Victorians were building their museums. But the nature of science has changed dramatically, certainly since World War II, and probably since the 1920s. We have more science in industry, in military research, in strategic research. So when you talk about science being communicated, what science are you talking about? Academic science is very different from industrial science. Academic science today in the Anglo-Saxon world is quite different from Germanic science in the 1860s or the 1880s.

Secondly, we have to be clear whether we are talking about the biomedical sciences or physical sciences or technological sciences in the universities. Historical and sociological work indicates clearly that the nature or process of research is quite different in particle physics, in solid-state physics, in biochemical research and so on.

Then there is the question of what can be called *Wissenschaft* but is not called science in the Anglo-Saxon world. Rather than getting bogged down in arguing about whether a creationist is a scientist you should try to be more coherent and systematic about what it is you are communicating, to whom, on whose behalf and with what consequences. Without an agenda like that we will end up going round in circles here.

Falk: Mac Laetsch was saying that unless we have a reasonable concept of what scientific literacy is, we are doomed to fail in our attempts to increase it. Most of us probably agree that people who are scientists, no matter how we define the term, have a better chance of rationally approaching the world than people who do not have that kind of training.

The issue still comes down to whether we can define science literacy. Is that even a useful term? Is it a useful goal to strive for in terms of the public? If it isn't, what kind of information should we be striving to provide for the public?

Friedman: I want to make a distinction between necessary and sufficient. We tend to argue that the kind of science communication we provide in our institutions is sufficient to improve the mechanisms of democracy in handling questions of science policy. Mac Laetsch has demonstrated that we have no evidence that this communication is sufficient. On the other hand I did not hear him attack the notion that it might be a *necessary*, if not a sufficient, component.

In the referendum on nuclear power in California a number of years ago an anti-nuclear power group made television commercials showing a bomb going off, implying that nuclear power plants can explode. A pro-nuclear power group commercial showed poverty-stricken people, implying that without this source of energy there was no economic basis for the survival of the state. Both sides assumed that no one would challenge the facts, which to an unfortunate extent was true. Without any kind of public science literacy the level of argument can degenerate to who produces the most emotionally persuasive commercial. On the other hand a perfectly scientifically literate community would still have difficulty deciding the issue of nuclear power.

I am accepting Mac Laetsch's argument that science literacy is not a sufficient condition for improving the functioning of democracy but I don't think he was arguing that it might not be necessary.

Dixon: People need to know more about the ideas of science. I think you said that the public ought to understand causality and probability as well as professional scientists do. But probability theory is philosophically and mathematically very complicated, and way above the level one needs for assessing day-by-day issues. For those issues people need to comprehend the general principle of probability, and such concepts as causality—whether A causes B or B causes A, or whether B and A are caused by C. It shouldn't be difficult for those concepts to be taught to everyone in school science courses. This is an important element of scientific literacy.

Whitley: That is elementary logic, isn't it? It doesn't go much beyond John Stuart Mill's canons of inductive logic. Is that science, in Laetsch's sense of the term?

Deehan: That is the demarcation dispute writ large, suggesting there is something special about science. But there isn't anything special about science in those kinds of day-to-day issues. Once you start thinking there is, you get into a lot of trouble.

Caravita: There seems to be a gap between the philosophy on which education in schools is based and the philosophy promoted by other institutions which have a concern for scientific literacy. In schools the purpose of education is not to build up an individual who can enjoy the environment and the world as best as possible and understand what is happening. School is mainly concerned with *instruction*, that is with the transmission of technical supports for enabling the students to read, to write, to calculate. In schools students are informed about the scientific interpretations of facts but they do not gain any further insight into the processes which underlie the interpretations, nor are they actively involved in such processes. This would be *education*. If we do not change the school educational policy, including the image of science conveyed by the teachers and textbooks, it will be very difficult to make the communication of science outside the school effective or to promote scientific literacy.

Laetsch: That leads us into how we communicate science. One of the reasons given for scientific literacy is that if you know the consequences of something, that knowledge will influence your behaviour. Yet quite a few studies show very disappointing results from great educational efforts in health. People may know a great deal but they still do not do what is good for them.

Wolpert: Not everybody has given up smoking but that does not mean that the information on smoking has had no effect.

Bodmer: The data on smoking incidence show a clear correlation with socioeconomic level and educational level. It is one of the best arguments for helping people in their health behaviour.

Laetsch: The correlation is for smoking by men.

Duensing: It applies to women too, but not so clearly. The scientific understanding of what smoking does helps to influence me, but it is not the only thing that influences me.

I think Mac Laetsch is saying that scientific literacy is important but it does not automatically make one a better decision-maker. There are other influences. In the AIDS crisis there were other factors that influenced the way people made a decision. It is not just knowing the information, it is also the process of how people gain that information.

Dixon: We are talking about independent variables but Walter Bodmer has pointed out that educational level *is* correlated with people's attention to their health. This is factual knowledge. It is necessary information but it is not sufficient to turn you into a healthy person.

Lucas: A lot of the arguments give the impression that scientific literacy is a sufficient condition. A lot of those justifications need to be attacked.

Wolpert: But who says those things?

Nelkin: The published studies on the effect of the press on changing people's behaviour and attitudes say that people read, interpret, absorb and act on scientific information on the basis of their predispositions. The predispositions, whether they come from social class influences or experience, affect the assimilation of scientific information (see literature review in Nelkin 1987).

Much of the discourse on scientific literacy in the USA is on computer literacy. It relates directly to the training of people in schools for jobs. Computer literacy has little to do with science.

Lucas: The Royal Society report on the public understanding of science (Bodmer et al 1985) does not say explicitly that scientific literacy is a sufficient condition, but that is the overall impression it gives.

Bodmer: If you can say that, you can't have read it carefully. The report says that one of the first things you need is a good general education—literacy and numeracy. It is self-evident that you have to put science literacy on top of a good general education.

Laetsch: I agree. The problem is that that is not how it is being proposed in the USA.

Bodmer: If you accept only the arguments for the humanistic case you will get about as much support for science and science education as we get for music and the arts in Britain. But is that enough?

Laetsch: That is a different issue. If we only go in that direction we will get to what was said earlier about computer science literacy. In many of our schools this is regarded as the equivalent of being scientifically literate, therefore other kinds of literacy are not being promoted in many schools in the USA, though these may be more important in the long run.

Falk: Once again we come back to whether the real concern at this point is the propaganda we are promulgating or our acting as if we really believe the propaganda. I guess what you are trying to say is that the latter is more concerning than the former.

Tisdell: Another problem is that we are all very limited as human beings and face terrible constraints in what we can learn and absorb. I am very illiterate in many areas of natural science. If I were to spend much time on it there would be terrific opportunity costs. Which areas are the priority areas for the public to learn about?

Wolpert: If you took out the word literacy from your presentation and used the word science would you say the same thing, Dr Laetsch?

Laetsch: The main arguments given in the USA (and elsewhere) for improving science education as well as science literacy are utilitarian arguments based on the factors I described. The National Science Board and the National Science Foundation have issued reports on this recently. The reports that we have had recently on pre-college education and the definitions of what every science museum in the country is about are all related to the strictly utilitarian benefits of being educated in and about science. We can call that scientific literacy, or just having a high level of science education, or whatever. This is the basis a lot of people use for their communication of science. In other words, if that is what you think science education is for, that is how you will communicate it. Those are political statements and they are tied directly to the source of funding. I am not arguing that science literacy is not a good thing or that to know science is not a good thing. I am saying that we need to examine whether the primary reason for science literacy is utilitarian, based on economic and political necessities.

Tan: In the western world the standard of general education is already at a certain level. In many parts of Asia, and elsewhere in the developing world, the first consideration is how to fill the rice bowl. In order to prove that something is good you must show results first.

Twenty-five years ago, the present major industrial zone in Singapore was a marshland and the economy depended on import/export and commodity trading. We had to convince about two million people that science and technology were going to affect their lives, because the government was going to industrialize the economy. Merely telling the general public that science is good for them will not convince them or gain an extra vote for the government.

Science education or scientific literacy is subservient to the immediate political ends or the economic ends but once people have their needs for housing and food met, they can see how these benefits were obtained. Then they are more likely to listen to new messages for so-called aesthetic reasons.

Laetsch: I agree with most of what you say but I have watched villagers involved in science activities in Asia, particularly in the public area. They don't differ from what one sees in science museums in London or elsewhere. If we understand why people are interested in science, whatever their level of education, that gives us a better understanding of how we might communicate science to people in many different institutions at many different levels, perhaps even within scientific disciplines. We don't always use the reasons people have for being interested in science as a basis for how to communicate science to them.

References

Bodmer WF, Artus RE, Attenborough D et al 1985 The public understanding of
    science. Royal Society, London
Nelkin D 1987 Selling science: how the press covers science and technology. W.H.
    Freeman, New York

Scientific literacy in the United States

JON D. MILLER

Public Opinion Laboratory, Northern Illinois University, DeKalb, Illinois 60115, USA

Abstract. Building on earlier work this paper describes a three-dimensional measure of scientific literacy. One dimension focuses on the understanding of the process or approach of science. A second dimension focuses on the existence of an adequate vocabulary for understanding scientific communication. The third dimension focuses on the relationship of science and technology to society. These three dimensions and a recently collected national survey indicate that about 5% of the adult population of the United States meets a minimal test of scientific literacy. This estimate is down from an earlier estimate of 7% in 1979. The implications of a high level of scientific illiteracy for the functioning of democratic government are discussed and it is recommended that more vigorous efforts be made to increase the level of scientific literacy.

1987 Communicating science to the public. Wiley, Chichester (Ciba Foundation Conference) p 19–40

The purpose of this essay is to provide a framework for a discussion of the communication of scientific information to the public. The concept of communication presupposes some level of literacy, even if only verbal or visual. The communication of scientific information presupposes some level of scientific literacy. There would be little disagreement with either of these propositions.

Scientific literacy, however, is one of those terms that is more often used than defined. In this paper, I will argue that scientific literacy is a necessary prerequisite for the communication of scientific information to the public. I will review an earlier estimate of the level of scientific literacy among adults in the United States and offer a new estimate of scientific literacy based on a national survey conducted a year ago. The paper will conclude with a discussion of the implications of low levels of scientific literacy for democratic government.

The evolution of the meaning of scientific literacy

Literacy has two distinct and quite different meanings. The older meaning of the term refers to being learned. The second meaning refers to the possession of minimal reading and writing skills and is the more common use of the term today. Unfortunately, much of the debate about scientific literacy has failed to distinguish between these two meanings of literacy.

Over the last century, a healthy debate has continued over whether a person can be considered learned without some knowledge of science and, conversely, whether a person with a knowledge of science might be considered to be learned without a thorough grounding in the traditional study of letters. The Rede Lectures at Cambridge have been the forum for several of the major contributions to this debate, beginning with Thomas Henry Huxley's lecture on 'Science and Culture', followed by Matthew Arnold's response in the 1882 lecture. C. P. Snow took up the argument again in his 1959 Rede lecture on 'The Two Cultures' and his argument was answered by F. R. Leavis in the Richmond Lecture of 1962. The debate was joined by Trilling (1962), Green (1962), Wollheim (1959), and others. The important point, however, is that all of these debates focused on the issue of the definition of being learned, not on the issue of communicating science to broader populations. In a *Daedalus* symposium, Levin (1965, p.2) characterized Snow's original lecture as an 'earnest plea for intercommunication across the high table, as between exponents of the scientific and humanistic disciplines'. Although the issues raised by Huxley, Arnold and Snow were important and are still with us, they are generally beyond the scope of this essay.

The second meaning of literate refers to the ability of an individual to read and write at a functional level, and we may extend that definition to suggest that scientific literacy in this context refers to the ability of an individual to read about, comprehend and express an opinion about scientific matters. The development of scientific literacy by a broader public did not become the subject of systematic study until the 1930s when Dewey (1934), in a paper entitled 'The Supreme Intellectual Obligation', declared that

> the responsibility of science cannot be fulfilled by methods that are chiefly concerned with self-perpetuation of specialized science to the neglect of influencing the much larger number to adopt into the very make-up of their minds those attitudes of open-mindedness, intellectual integrity, observation, and interest in testing their opinions and beliefs, that are characteristic of the scientific attitude.

After Dewey's charge, a number of science educators began to think about the formal definition and measurement of the scientific attitude. I. C. Davis (1935, p.119) defined the meaning of scientific attitude:

> We can say that an individual who has a scientific attitude will (1) show a willingness to change his opinion on the basis of new evidence; (2) will search for the whole truth without prejudice; (3) will have a concept of cause and effect relationships; (4) will make a habit of basing judgment on fact; and (5) will have the ability to distinguish between fact and theory.

Noll (1935) and Hoff (1936) offered similar definitions and began the task of developing items for use in testing. Virtually all of the empirical work before the Second World War focused on the development of a scientific attitude.

With the post-war growth in standardized testing, a number of science educators and test developers began to focus on the level of comprehension of basic scientific constructs and terms. Epitomized by the standardized tests of the Educational Testing Service and the College Board, a growing number of studies attempted to chart the level of cognitive scientific knowledge among various groups in the population.

Beginning in the mid-1960s, the National Assessment of Educational Progress (NAEP) began to collect data concerning the level of scientific and other categories of knowledge from national random samples of precollegiate students. The NAEP studies are particularly noteworthy since they are the only national data collection programme that eliminates the bias of self-selection inherent in voluntary testing programmes. The NAEP studies were also the first to systematically include measures of an understanding of the norms or process of science as well as of the cognitive content of the major disciplines.

The combination of these two dimensions—an understanding of the norms of science and knowledge of major scientific constructs—may be viewed as the traditional meaning of scientific literacy as applied to broader populations.

In the last 15 years there has been a growing concern about the level of knowledge that the public holds about various public policy issues that are scientific or technological in character. This concern was fed by several rather different streams. On the one hand, a number of environmental conservation groups began to find that it was necessary for individuals to have some minimal scientific knowledge to follow the debates about the pollution of the environment. A number of new science education courses were developed and introduced in response to this concern. More recently, the increasing number of state referenda on issues like nuclear power and laetrile have generated concern in the scientific community about the ability of the public to understand the issues and make an informed judgement. Similar concerns were expressed in the 1950s when numerous localities held referenda on the fluoridation issue and when anti-fluoridation groups won several of these referenda.

Morrison (1969, p.156) summarized the case for a broader public understanding of public policy issues involving science:

Science can no longer be content to present itself as an activity independent of the rest of society, governed by its own rules and directed by the inner dynamics of its own processes. Too many of these processes have effects which, though beneficial in many respects, often strike the average man as a threat to his autonomy. Too often science seems to be thrusting society as a whole in directions in which it does not fully understand and which it certainly has not chosen.

The scientific community must redouble its efforts to present science—in the classroom, in the public press, and through education–extension activities of various kinds—as a fully understandable process, 'justifiable to man,' and controllable by him.

Shen (1975) has characterized this dimension as 'civic science literacy'.

In general, I would argue that the meaning of scientific literacy as applied to general populations has evolved from a two-dimensional concept to a three-dimensional concept, reflecting the need for public awareness of the impact of science and technology on society and the inherent policy choices that emerge.

Some previous empirical measures of scientific literacy

Although the level of scientific literacy has been the subject of lively discourse for at least 50 years, there have been surprisingly few efforts to provide empirical measures of scientific literacy for the population in general, or even for major segments of it. Until the last decade, nearly all of the major efforts to measure scientific literacy focused on populations of school age, and most often on small purposely selected sets of schools or classrooms. For a nation generally disposed toward social measurement this is a disappointing database.

As a preface to an analysis of the current levels and distribution of scientific literacy, I will review the previous efforts at empirical measurement in relation to the three dimensions outlined above.

The norms and methods of science

Most of the early empirical work focused on the definition and measurement of the scientific attitude, as described by Dewey (1934). I. C. Davis (1935) sought to develop sets of items that would use examples from the everyday experience of the student and create situations which would force a judgement. By providing a set of alternatives that reflected rational, hunch or superstitious alternatives, Davis attempted to measure the student's utilization of the scientific attitude. For example, students were presented with statements like 'Red hair means that a person has a fiery temperament', or 'A disease is a punishment for some particular moral wrong', or 'Air is composed of molecules'. Davis (1935, p.120) collected data from about 1000 Wisconsin high-school students, and reached the following interesting conclusions:

(1) High-school pupils in Wisconsin are not superstitious.
(2) High-school pupils make almost as good records as the teachers.
(3) Many of the theories in science are being taught as facts by many of our best teachers.
(4) Pupils seem to have a fairly clear concept of the cause-and-effect relationship, but they do not seem to be able to recognize the adequacy of a supposed cause to produce the given result.
(5) Many teachers tend to propagandize their material when there is no scientific evidence for the statements they make.

(6) Teachers do not consciously attempt to develop the characteristics of a scientific attitude. If pupils have acquired these characteristics, it has come about by some process of thinking or experiences outside of the science classroom.

Similar test development was also undertaken by Noll (1935) and Hoff (1936), stressing the use of everyday examples to measure the use of scientific thinking.

The effort to provide a sound empirical measure of adherence to scientific thinking continued throughout the post-war years. Schwirian (1968) used factor analysis to develop a five-dimensional measure of scientific thinking. The five dimensions—rationality, utilitarianism, universalism, individualism, and a belief in progress and meliorism—were patterned after Barber's (1962) analysis of science and the social order. The Schwirian scale was originally tested on samples of undergraduates from a midwestern university and has since been used by other researchers on a number of local samples, but it has yet to be used on either a national or a broad population sample.

The National Assessment of Educational Progress has included in its science testing programme a number of items concerning knowledge of scientific norms and the ability to reason to conclusions from relatively simple data. These studies provide the only national estimates of the level of knowledge about scientific norms or about the ability of young people to think in logical and ordered terms.

In NAEP's 1972–73 assessment of science achievement, students aged 9, 13 and 17 years were asked to perform a set of simple experiments and to explain the results. For example, each student was shown unlabelled pieces of sandstone, quartz and granite and asked to select the stone 'most likely formed under water.' The sandstone sample was selected by 60% of 9-year-olds, 71% of 13-year-olds, and 78% of 17-year-olds. However, when asked to explain why they had selected the sandstone sample, only 7% of 9-year-olds, 15% of 13-year-olds, and 26% of 17-year-olds were able to describe the process of sedimentation (even vaguely) or to provide other acceptable explanations (NAEP 1975). Similar patterns were found in several other experiments involving colour mixing, temperature mixing, relative volume, rotation and revolution, and simple circuit boards. These results suggest that much of what may appear to be scientific knowledge in multiple-choice testing is not supported by an *understanding* of the underlying scientific principles and processes. Although numerous attempts have been made to measure what Dewey (1934) called the 'scientific attitude' among school-aged or young adult groups, the first national survey to measure adult comprehension of the scientific process was the 1957 science news study. In that study, each respondent was asked to define the meaning of scientific study, and the open-ended response was coded into a set of categories that reflected various perceptions of scientific study. Withey (1959) concluded that only about 12% could be said to have a reasonable understanding of the term.

A 1979 survey (Miller & Prewitt 1979) included the same question, asking each respondent to assess whether he or she had a clear understanding of the meaning of scientific study, a general sense of its meaning, or little understanding of its meaning. Those persons declaring that they had a clear understanding of the term were asked to explain its meaning in their own words, and the interviewer recorded the response verbatim[1]. The responses were later coded by two coders independently, and cases involving disagreements in coding were judged by the coding supervisor.

The resulting data indicated that about 14% of the adults in the United States in 1979 were able to provide a minimally acceptable definition[2] of the meaning of scientific study (Miller et al 1980a, Miller 1983b). This level of understanding was not statistically different from the 12% in 1957 reported by Withey (1959), suggesting that there was little improvement in the proportion of American adults correctly understanding the meaning of scientific study in the first two decades after Sputnik.

Following the multiple measurement principle, the 1979 survey also included a short battery of items about astrology. The survey asked each respondent about the frequency of his or her reading of astrology reports, and then asked the respondent to characterize astrology as very scientific, sort of scientific, or not at all scientific. While only 8% thought that astrology was very scientific, an additional 34% thought that it was 'sort of scientific'. The other half of the respondents recognized that astrology is not scientific. In general, these responses provide a good check on the previous item about the meaning of scientific study.

Although a respondent may have been able to recite an acceptable definition of scientific study, a correct understanding of the process should have led to a rejection of astrology as scientific. Following this logic, the 1979 measure

---

[1] It is important to acknowledge some of the limitations inherent in the measurement process. The questions concerning the understanding of scientific study illustrate the limitations and utility of survey measurements. The query as to whether an individual has a clear understanding of the meaning of scientific study is inherently subjective. Undoubtedly, some individuals overestimated their level of understanding, while others underestimated it. Yet, the subjective nature of the response is also its strength, since a person who does not think that he or she understands the term would be unlikely to utilize it in communicating with others. The pairing of this subjective query with the more objective and substantive follow-up definition and astrology questions allowed for a correction for overestimates of understanding in the subjective query. The resulting estimate of the proportion of adults with an understanding of the meaning of scientific study should be taken as an upper limit on the occurrence of this understanding under more rigorous measurement methods.

[2] To be classified as understanding the process of scientific study, a respondent had to report that scientific study involved (1) the advancement and potential falsification of generalizations and hypotheses, leading to the creation of theory, (2) the investigation of a subject with an open mind and a willingness to consider all evidence in determining results, or (3) the use of experimental or similar methods of controlled comparison or systematic observation. Responses that characterized scientific study as the accumulation of facts, the use of specific instruments (i.e. looking at things through a microscope), or simply careful study were coded as incorrect.

of understanding of scientific study (Miller & Prewitt 1979) required a respondent to be able to provide a reasonable definition of scientific study *and* to recognize that astrology is not scientific. Only 9% of adult Americans met this rather minimal test (Miller 1983b).

Cognitive science knowledge

The second dimension of scientific literacy is an understanding of basic scientific constructs. The argument here is simple and clear. If an individual cannot comprehend basic terms like atom, molecule, cell, gravity or radiation, then it would be nearly impossible for that person to follow much of the public discussion of scientific results or of public policy issues pertaining to science and technology. In short, a minimal scientific vocabulary is necessary if one is to be scientifically literate.

As the use of standardized testing expanded during the 1950s and 1960s, a number of tests were developed to measure a student's knowledge of basic scientific constructs (Buros 1965). The majority of these tests have been used by teachers and school systems to evaluate individual students, to determine admission or placement, or for related academic counselling purposes. While some test-score summaries have been published by the Educational Testing Service and other national testing services, these reports reflect only those students who plan to attend college or who have elected to take the test for some reason. Although very large numbers of tests are used each year, the self-selected nature of the student populations involved continues to raise substantial problems for interpretation and analysis.

The only national data set that provides scores of cognitive science knowledge for broad and randomly selected populations is the National Assessment. For over a decade, the periodic tests of the NAEP have collected cognitive science knowledge data from national samples of 9-, 13-, and 17-year-olds. In some years, a national sample of young adults aged 26–35 was also used to assess the continuing impact of formal study.

On the basis of three assessments between 1969 and 1977, the National Assessment found declining science achievement scores for all age groups and almost all socioeconomic subgroups. Female students, black students, students whose parents did not complete high school and students who live in large central cities are all substantially below the national average in science achievement (NAEP 1978a,b).

Our 1979 survey (Miller & Prewitt 1979) included three items that may be said to represent basic science and social science terminology. Each survey respondent was asked to assess their own understanding of radiation, DNA and GNP. As with the meaning of scientific study, each respondent was asked to report whether he or she held a clear understanding of each term, a general sense of its meaning, or little understanding of its meaning. Although no probes were made to check the accuracy of the self-reported assessment, the experience of the respondent with the previous probe in regard to the meaning of scientific study should have discouraged optimistic reporting of knowledge levels[3].

The 1979 study found that about half of the adult population of the United States believed that they had a clear understanding of radiation, about a third thought that they understood GNP, and only one in five asserted a knowledge of the meaning of DNA. To provide a single measure of this dimension, a summary index was constructed. To be classified as knowledgeable about basic scientific constructs, a respondent was required to report a clear understanding of at least one of the three constructs and not less than a general sense of a second term. The 1979 data indicated that exactly half of the adult population met this minimal requirement (see Miller 1983b).

Attitudes toward organized science

The third dimension of scientific literacy involves an understanding of some of the major public policy issues that involve or directly affect the conduct of science and technology. As science becomes increasingly dependent on public support and as public regulation reaches deeper into the conduct of organized science, the frequency and importance of science policy issues on the national agenda will undoubtedly increase. Shen (1975) has estimated that slightly over half of the bills introduced in the Congress involve science or technology in some degree. The establishment of a standing Committee on Science and Technology in the United States House of Representatives reflects the growing volume and importance of scientific and technological issues in the national political system.

The first national study that included a meaningful set of measures of attitudes toward and knowledge about organized science was the 1957 National Association of Science Writers (NASW) study (R. C. Davis 1958, Withey 1959). In general, the survey found that only a minority of the public displayed a strong interest in scientific issues and that the level of public knowledge about science was low. The public tended to hold high expectations for the future achievements of science and technology, but displayed some awareness of the two-edged nature of the scientific enterprise.

Beginning in 1972, the National Science Board initiated a biennial survey of public attitudes toward science and technology, reported in the *Science Indicator* series (NSB, 1973, 1975, 1977, 1981, 1983, 1986). These surveys

---

[3] As discussed in regard to the meaning of scientific study questions, it is important to note the limitations inherent in these measures. If probes had been employed, some of the respondents who rated their understanding of these terms to be high would have been re-classified downward. These estimates should be taken as upper limits of the true value in the population.

generally found that the public retained a high level of appreciation and expectation, but there were some signs of an increasing wariness in the public. There was, however, no evidence to support the idea that a strong anti-science sentiment was developing in the United States.

The 1979 NSB *Science Indicators* survey included a useful measure of this dimension of scientific literacy. The survey asked a battery of items about three controversies—the use of chemical additives in food, nuclear power and space exploration. For each of these three areas, each respondent was asked to cite two potential benefits and two potential harms. About two-thirds of the 1979 respondents were able to cite at least two benefits or dangers (or one of each) for food additives and nuclear power, but only a third were able to display a similar level of information about the potential benefits and harms of space exploration.

To provide a single measure of this dimension of scientific literacy, a respondent was scored as having a minimally acceptable level of information on the issues if he or she was able to name six potential benefits or harms out of a possible 12 inquiries and probes. On this criterion, about 41% of adult Americans qualified as knowledgeable about scientific and technological public policy issues.

Beginning in 1979, the focus of the NSB *Science Indicators* series changed from general attitudes to a more structural approach that attempted to identify that segment of the population that holds some level of interest in and knowledge about science policy matters and to examine the attitudes of this 'attentive public' (Miller et al 1980a, NSB, 1981, 1986, Miller 1983a). The results indicated that about one in five American adults follow scientific matters on a regular basis and that this attentive group was generally more favourable to organized science than the general population groups previously measured.

There is also evidence to suggest that the attitudes of young adults toward organized science are even more positive than those of their elders. In a 1978 national survey of high-school and college students (Miller et al 1980b), we were able to identify the developmental roots of adult attentiveness to science matters and to determine that the attitudes of these young adults were generally positive and supportive for science. As with their elders, substantial segments of the young adults in the 1978 study displayed low interest in and had little information about scientific and technological matters. About 90% of high-school students not planning to attend college failed to meet minimal criteria for interest in scientific issues or cognitive knowledge of basic scientific constructs. These results do not support the idea that the next generation will be anti-scientific, but rather it may be expected to follow the general contours of current adult attitudes, adjusted for the level of educational achievement of the new generation.

A 1979 estimate of scientific literacy

The preceding review of measures of the three dimensions of scientific literacy suggests that the efforts have been relatively uncoordinated and that there
has been little consensus on a comprehensive definition of scientific literacy.
Although a number of partial measures had been developed, the first consoli-
dated measure of the proportion of the American people who were scientifically literate was the estimate made in 1979 (Miller 1983b). This measure of
scientific literacy required respondents to demonstrate a minimal understanding of the scientific process, a set of scientific terms, and the societal impact of
science and technology if they were to be classified as scientifically literate.

The response to the 1979 measure indicated that only 7% of the adult
population qualified as scientifically literate (Miller 1983b). When it is recal-
led that each percentage point in a national survey represents about 1.7
million adults, this result would indicate that about 12 million American
adults were scientifically literate in 1979. The 1979 finding was discouraging
to many persons who had devoted considerable effort to encouraging broader
public understanding of science and technology, but it was not unexpected by
many scholars within the scientific community. No one has challenged the
general accuracy of the estimate.

The 1985 level and distribution of scientific literacy

In the years since the responses to the 1979 measure were collected, numerous groups and scholars have wondered whether the level of scientific literacy
has been changing in the United States. With the growth in the readership of
science magazines and the viewership of science television, it has been suggested that the level of scientific literacy might be expected to increase
gradually. I have made that suggestion from time to time. One of the purposes of this paper is to present a new estimate of scientific literacy, based on
a set of responses to almost identical measures collected in November and
December 1985.

The 1985 study incorporated the same questions on the understanding of
the process of science as were used in 1979. Four independent coders assessed
the responses and the results indicated that 13% of American adults could
provide a minimally acceptable explanation of the process or approach of
science. This result is comparable to the 12% found in 1957 and the 14%
reported in 1979. Substantively, it appears that the level of public understanding of what it means to study something scientifically has not changed significantly over the last three decades.

Following the 1979 procedure, we adjusted this estimate by removing
respondents who indicated substantial misunderstandings of the scientific

process in other areas. Specifically, as in 1979, those respondents who indicated that astrology was either 'sort of' or 'very' scientific were dropped from the classification of those who understand science. In the 1985 study, this correction reduced the proportion of adults with a correct understanding to 9% of the adult population.

The second dimension of scientific literacy—the understanding of basic terms and concepts—was measured in 1985 with the identical set of three items used in 1979. The results indicated that significantly fewer American adults thought they had a clear understanding of radiation, DNA or GNP in 1985 than the number reported in 1979. The proportion of adults who thought they had a clear understanding of radiation dropped from 49% in 1979 to 29% in 1985. The proportion with a clear understanding of DNA dropped from 22 to 14%, and the proportion believing they have a clear understanding of GNP dropped from 31 to 23%. A special analysis of these results by demographic characteristics found a similar pattern of reduced confidence in understanding in all educational, age and gender groups.

When the results of the separate answers were combined into a single index—following the same procedure used in 1979—the proportion of American adults scoring high on this dimension dropped from 50% to 34%. What can explain this sharp decline? Let me suggest one substantive explanation.

As the volume of scientific and technical information available to the public increases, each individual has more opportunities to receive and try to understand communications that include basic scientific terms. This may involve reading a science-related story in the newspaper or watching a *Nova* programme on television. To the extent that an individual has difficulty in understanding or being able to use those terms, there may emerge an increased sense of the difficulty of using those terms. The increased flow of scientific and technical information into the media mainstream may have stimulated an adjustment in the self-reported level of understanding.

The third and final dimension of the 1979 scientific literacy measure focused on the understanding of the impact of science and technology on society. The 1979 study asked each respondent to name the advantages and disadvantages of three scientific or technological activities. Those measures were not repeated in 1985, but a comparable set of items were used in an agree–disagree format. A number of special analyses indicate that the two sets of measures produce comparable results.

In the 1985 study, each respondent was asked to name the most important achievement of science and technology in the last 10 years and the second most important achievement. For the purpose of this dimension, each respondent earned one point if they could name any such achievement and a second point for naming a second achievement. As in 1979, these responses were not evaluated in terms of correctness.

In addition to the two open-ended questions, the 1985 study included a set of statements that were read to each respondent, who was asked to agree or disagree with each statement. Each respondent would earn up to five additional points by agreeing that smoking causes serious health problems, agreeing that there are probably thousands of planets like our own in the universe on which life could have developed, agreeing that humans developed from earlier species of animals, disagreeing with the idea that rockets and space shots have changed our weather, and disagreeing that some of the unidentified flying objects reported have been space vehicles from other civilizations. All of these statements reflect points of contact between science and public policy. Those respondents with a score of four or more on this seven-point index—67% of adults—were classified as having a higher level of understanding of the relationship between science and society.

The combination of the three dimensions just described provides a new estimate of the level of scientific literacy in the United States. Using the same rule as in 1979, the results indicate that only 5% of American adults qualified as scientifically literate in 1985 (see Table 1). This is a statistically significant decline (at the 0.05 level), but it is not substantively important. The essential point is that the level of scientific literacy in the United States remains low and that the informal science education efforts of recent years have not produced any measurable increase in scientific literacy.

Implications for a democratic society

If the estimate that only 5% of the adult population is scientifically literate is correct, what are the implications for our society in general and our political system in particular? The implications are obviously broad, ranging from the quality of the cultural experience of people in our society to the ability of consumers to judge among competing products in the marketplace. Other contributors to this symposium will discuss many of the cultural and economic implications of these results, but I will focus my attention on the implications of the low level of scientific literacy for the formulation of science policy in a democratic political system.

If only 5% of the adult population is scientifically literate, how is science policy to be made? Can the basic tenets of a participatory political system be maintained? To address these issues, it is necessary to put these results in the context of a more general process of political specialization that has been occurring in the American political system in recent decades.

Political specialization

Political specialization refers to a process in which individuals (1) decide whether or not to follow political affairs at all, and (2) select the issue or issues for which they are willing to invest the time and other resources

TABLE 1   The percentage of the public classified as scientifically literate

	Year	Scientifically literate (%)	N
All adults	1979	7	1635
	1985	5	2005
Age			
17–34	1979	11	670
	1985	7	800
35–54	1979	7	492
	1985	5	614
55 and over	1979	3	473
	1985	2	591
Gender			
Female	1979	6	867
	1985	3	1054
Male	1979	9	773
	1985	6	950
Education			
Less than HS	1979	0	465
	1985	0	512
HS graduate	1979	2	550
	1985	3	1006
Some college	1979	14	382
	1985	7	139
College graduate	1979	22	146
	1985	12	229
Graduate degree	1979	26	92
	1985	18	121

HS, high school.

necessary to become and remain informed. These are not clearly delineated decisions made at a particular time but rather reflect the outcome of continuing competing demands on time and attention. The era is long past when any individual could hope to acquire and maintain mastery of more than a few political issues at any given time. Unlike our frontier ancestors who waited eagerly for old newspapers and magazines from the East, the modern citizen can only sample selectively from the avalanche of information routinely available.

The need for specialization springs from a combination of two basic forces. First, citizens in the latter part of the 20th century are faced with a large number of competing demands for their time, while time remains finite. There is ample evidence of the growth of time pressures over the last several decades (see Szalai 1973). Fewer adults than formerly choose to devote a large share of time to political affairs and thus we have observed a steady decline in public participation in the political system over the last four decades. Even presidential elections, which command the highest levels of public concern and participation, attract the votes of barely half of the eligible adults in the United States.

Second, during the same period of time, the information threshold for many political issues has been increasing. It now requires more specialized information to be knowledgeable about almost any given political issue. Issues involving science fall into this category, as do most of the issues on the national political agenda. Even an area like economic policy—often referred to in traditional political jargon as the pocketbook issue—has become increasingly complex and is beyond the command of a substantial majority of American adults. If only one in four adults have a clear understanding of a term like 'GNP', how many citizens might be expected to comprehend a debate over the position of the dollar, the pound or the yen in international monetary markets?

The combined effect of these two forces has been the narrowing of the political horizon of most American adults to only a small number of political issues. It is important to note that this is not the same process as 'single issue' politics that revolve totally around one strongly felt political position, but rather a rational and gradual narrowing of the number of topics on which an individual can hope to remain adequately and currently informed. Generally, the scope of interest in various issues has exceeded the scope of information held by most people.

How, then, does this specialization process affect the more general political system? Three decades ago, Almond described the process as it applied to public attitudes toward and public participation in foreign policy. In his original work, Almond (1950) outlined a pyramidal structure that illustrated the types of public participation in the political process that were likely to occur under conditions of issue specialization (see Fig. 1). In this stratified model, the decision-makers are at the pinnacle of the system and represent those persons who have the power to make binding policy decisions. This group would include a mix of executive, legislative and judicial officers. In science policy, the officers would be primarily at the federal level.

The second level of the model is the group of non-governmental policy leaders that are often referred to as elites in political science. This leadership group interacts regularly with decision-makers, and Rosenau (1961, 1963) and others have noted that there is some flow of elites into decision-making posts and of decision-makers into the leadership group from time to time. When there is a high level of concurrence between the decision-makers and the leadership group, policy is made and normally there is no wider public participation in the policy process.

On some issues, however, views within the policy leadership group may be divided; in that circumstance, there may be appeals to the attentive public to

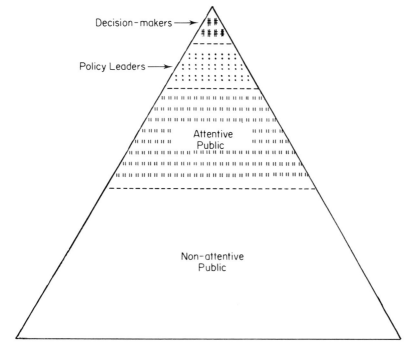

FIG. 1   A stratified model of science and technology policy formulation.

join in the policy process and to seek to influence decision-makers by direct contact and by persuasion. The attentive public—the third level of the model— is composed of individuals who are interested in a particular policy area and willing to become and remain knowledgeable about the issue. In 1979, the attentive public for science policy included about 27 million adults, or about 18% of the adult population (Miller et al 1980a, NSB 1981, Miller & Prewitt 1982).

Although the processes through which the leadership group mobilizes the interest of the attentive public in science policy are only now being studied, it would appear that there is a flow of appeals from leaders to the attentive members of the public through professional organizations, through specialized journals and magazines, and through employment-related institutions. These attentive individuals probably contact public officials—decision-makers—by the traditional avenues of letter writing, telephone contacts and personal contacts.

A second circumstance in which attentive individuals become involved is when there is a high degree of consensus among the leadership group but a lack of concurrence by some or all of the decision-makers. In this case, the leadership groups work to mobilize the attentive public to make contacts with decision-makers and to argue directly for the preferred policy. The recent discussions of federal funding for science illustrate this case.

At the bottom of the pyramid is the general, or non-attentive, public. For the most part, these individuals display a low level of interest in science policy and low levels of knowledge about organized science. It is important, however, to understand two points about this group. First, the general population always retains a political veto if they should become sufficiently unhappy about the policies that the decision-makers, leaders and attentive individuals have fostered in any area. The role of the public in ending the wars in Korea and Vietnam illustrates the operation of the popular veto power. It is this ability to intervene and veto that sustains the democratic nature of the policy formulation process.

Second, it is important not to equate non-attentiveness to science policy with ignorance or lack of intellectual activity. We are all non-attentive to a vast number of issues. Many of those in the non-attentive group for science policy may be well educated, interested in and knowledgeable about other issues, and politically active. The 1985 data indicated that about half of the adult population in the United States were attentive to one or more political issues.

Scientific literacy and the formulation of science policy

Given the stratified model outlined above, the low level of scientific literacy raises two important problems.

First, the level of scientific literacy among the 80% of the population not attentive to science policy is extraordinarily low—in the 2–3% range. In the context of the political specialization process, we would not expect a very high rate of scientific literacy in the non-attentive public and few voters would be expected to make their choices primarily on the basis of scientific or technological issues. In recent years, however, an increasing number of referenda have concerned issues related to science or technology—nuclear power, laetrile, recombinant DNA facilities, fluoridation—and the low level of scientific literacy argues strongly that a substantial majority of the general electorate are ill-prepared to make those judgements.

Second, over 90% of the attentive public for science policy did not meet the minimal criteria for scientific literacy in 1985. This is a surprising result, but not an inexplicable one. About a third of the adult population displayed a minimal facility with basic scientific terms and 60% per cent held at least a minimal level of information on the concepts tested. Many of these individuals may have had a high level of interest in one or more science-related issues and may have followed those issues through popular magazines like *National Geographic* or *Discover*, but still not understood the scientific process. We might expect this group to have more difficulty with articles in *Science*, *Scientific American*, or a professional or disciplinary journal, but they may be able to write very clear letters to their elected representatives.

This group would appear to be dependent on science journalists to 'interpret' the scientific debate on most issues.

As more appeals are made to the attentive public to pressure directly for public policy outcomes, the importance of this group will grow. Yet, given the large proportion of this group that are dependent on 'translators', it is a fragile situation. The personality or philosophical perspective of the translator may become as important as the substance of the scientific arguments.

Ideally, the scientifically literate proportions of both the general and the attentive publics should be increased. There are compelling cultural, economic and political arguments for a major effort to expand scientific literacy. The most effective point of intervention for increasing the scientific literacy of the broader population is the improvement and enrichment of pre-collegiate education. Beyond improved science education at the pre-collegiate level, I would argue that high priority should be placed on the expansion of scientific literacy among the attentive public for science policy. To a large extent, the political specialization process assumes a basic level of literacy in any given area among attentive individuals and the process operates most effectively under that condition. This is an interested audience searching for more sophisticated information than is normally available in the news media. The growth of semi-sophisticated science magazines like *Discover* reflects the desire of the attentive public for science policy for more and better information.

From Dewey to Morrison, the scientific community has been urged to seek to communicate more effectively and openly with the public, especially the informed public. There is some evidence that the scientific community is beginning to respond to this challenge and that there is a new level of willingness on the part of scientists and engineers to seek to explain their problems and aspirations to interested lay audiences. If this communication is to continue and expand, however, there must be an audience that is capable of understanding both the substance of the arguments and the basic processes of science. If the science policy process is to function effectively, the educational needs of the attentive public for science policy must be addressed at an early date.

References

Almond G 1950 The American people and foreign policy. Harcourt Brace, New York
Arnold M 1882 Literature and science. Nineteenth Century 12:216–230
Barber B 1962 Science and the social order. Free Press, New York
Buros OK 1965 The sixth mental measurements yearbook. Gryphon Press, Highland Park, New Jersey
Davis IC 1935 The measurement of scientific attitudes. Science Education 19:117–122
Davis RC 1958 The public impact of science in the mass media. Survey Research Center, University of Michigan (Monograph No. 25)
Dewey J 1934 The supreme intellectual obligation. Science Education 18:1–4

Green M 1962 A literary defense of 'The two cultures.' Kenyon Review, p 731–739

Hoff AG 1936 A test for scientific attitude. School Science and Mathematics 36:763–770

Huxley TH 1898 Science and culture. In: Collected essays. Appleton & Co., New York

Leavis FR 1963 Two cultures? The significance of C.P. Snow. Pantheon Books, New York

Levin H 1965 Semantics of culture. In: Holten G (ed) Science and culture. Beacon Press, Boston

Miller JD 1983a The American people and science policy. Pergamon, New York

Miller JD 1983b Scientific literacy: a conceptual and empirical review. Daedalus 112(2):29–48

Miller JD, Prewitt K 1982 A national survey of the non-governmental leadership of American science and technology. A report to the National Science Foundation under NSF grant 8105662. Public Opinion Laboratory, DeKalb, Illinois

Miller JD, Prewitt K, Pearson R 1980a The attitudes of the U.S. public toward science and technology. A final report to the National Science Foundation

Miller JD, Suchner RW, Voelker AV 1980b Citizenship in an age of science. Pergamon, New York

Morrison RS 1969 Science and social attitudes. Science 165:150–156

National Assessment of Educational Progress 1975 Selected results for the national assessments of science: scientific principles and procedures. Science Report No. 04-S-02. Educational Commission of the States, Denver, Colorado

National Assessment of Educational Progress 1978a Three national assessments of science: 1969–1977. Science Report No. 08-S-00. Educational Commission of the States, Denver, Colorado

National Assessment of Educational Progress 1978b Science achievement in the schools. Science Report No. 08-S-01. Educational Commission of the States, Denver, Colorado

National Science Board 1973 Science indicators 1972. U.S. Government Printing Office, Washington DC

National Science Board 1975 Science indicators 1974. U.S. Government Printing Office, Washington DC

National Science Board 1977 Science indicators 1976. U.S. Government Printing Office, Washington DC

National Science Board 1981 Science indicators 1980. U.S. Government Printing Office, Washington DC

National Science Board 1983 Science indicators 1982. U.S. Government Printing Office, Washington DC

National Science Board 1986 Science indicators 1985. U.S. Government Printing Office, Washington DC

Noll VH 1935 Measuring the scientific attitude. Journal of Abnormal and Social Psychology 30:145–154

Rosenau J 1961 Public opinion and foreign policy: an operational formulation. Random House, New York

Rosenau J 1963 National leadership and foreign policy: the mobilization of public support. Princeton University Press, Princeton, New Jersey

Schwirian PM 1968 On measuring attitudes toward science. Science Education 52:172–179

Shen B 1975 Scientific literacy and the public understanding of science. In: Day S (ed) Communication of scientific information. Karger, Basel

Snow CP 1959 The two cultures and the scientific revolution. Cambridge University Press, New York

Szalai A 1973 The use of time. Mouton, The Hague

Trilling L 1962 Science, literature & culture: a comment on the Leavis–Snow controversy. Commentary, p 461–477

Withey SB 1959 Public opinion about science and the scientist. Public Opinion Quarterly 23:382–388

Wollheim R 1959 Grounds for approval. The Spectator, p 168–169

DISCUSSION

Nelkin: Do you include religion in your surveys? Religious variables are important, at least in the attitudinal work that has been done in the United States.

Miller: We had more religious variables in our recent survey than in 1979. Religious fundamentalism in the United States tends to be inversely related to formal education, while an interest in science and being literate in science are highly correlated with formal education. We need to find out how much of the difference in scientific literacy is due to fundamentalism and how much simply reflects a poor education that produces a lack of interest in science, or hostility to it.

Nelkin: The influence of religious fundamentalism on education is beginning to change in the United States.

Miller: But there is a lot of religious conservatism among college students, 47% of whom live within 50 miles of home. They go home three out of four weekends and still go to their parents' church. The familial string is not being broken.

Macdonald-Ross: What proportion of the population in the student age group major in science?

Miller: About half of the high-school graduates in the USA go to university. About 10% of those major in science or engineering, I think.

Macdonald-Ross: In the population you investigated, with 5% of scientifically literate individuals, how many graduated with science as a major subject?

Miller: Almost all had two years of college-level science courses, even if they majored in another subject. High-school science and scientific literacy are not closely correlated. An appropriate national goal would be to make high-school graduates scientifically literate, but that would cost twice as much and probably take twice as long as the Apollo programme.

Bodmer: What proportion of those who majored in science were scientifically literate?

Miller: Less than 100%.

Bodmer: In a 2×2 table what would the correlation be between your three dimensions?

Miller: The defining variable is understanding the process of science. At that point, the number is down to 10 or 12% and the others are in the order of 30–50%.

Bodmer: So about half of those who have an understanding of science do not match your other criteria of knowing what DNA, GNP and radiation are.

Lucas: What proportion of your scientifically literate respondents are in the attentive group?

Miller: Ninety per cent. The problem is that only about 10 or 12% of the attentive people are scientifically literate.

Caravita: How do you define 'attentive'?

Miller: In 1979, we used 32 headlines about new scientific discoveries, economic policy, foreign policy, or sports events and asked people which stories they would have read after seeing the headlines. When we asked whether the respondents were very interested, moderately interested or not very interested in these different areas, there was a correlation of about 0.9 with the factors in the reading pattern. Since the correlation was so high in 1979 we didn't use headlines in our recent study. Instead, we asked which areas of news people were interested in, among a list that included foreign policy, agricultural policy, economic policy, some science issues and education issues, and so on. There were enough items on the list for most people to find one or two areas in which they were interested.

Caravita: How do you know whether they transform this interest into action?

Miller: We do several things. We ask them whether they think they are very well informed, moderately well informed or not well informed about each item on the list. The attentive people are those who say they are very well informed and very interested. That is an important intersection. We also ask the whole sample if they have contact with their legislators or contribute to campaigns or any of a variety of political activities. I have shown (Miller 1983) that these people are quite active politically, but not 100% active. The problem is to get those who are attentive to do something. The science community in the United States has probably been able to mobilize 2%, at most, of the attentive population, but it is a waste of time trying to mobilize those at the bottom of the pyramid to write letters and so on. It is more sensible to find those who are already interested in a topic and try to mobilize them.

Hearn: What are the ways in which the 5% who are scientifically literate maintain their interest in science?

Miller: They consume a lot of information—they watch television, they read magazines, they go to museums. What distinguishes them from the rest of the population is that they consume a lot of print. A lot of people have lost the ability to read and understand moderately complex material. This ultimately erodes their depth of understanding so that they become more superficial and less informed about things. The museum or television experiences are stimuli that prod people to find out more by buying a book. The stimulus makes people make an effort to consume more information in depth.

Dixon: Your leaders are people who set the agenda for public debate. Can you tell us anything about particular subgroups such as newspaper editors, who are the real gatekeepers, rather than the specialist science correspondents in the media?

Miller: We have interviewed three or four samples of 500 people. Editors and science writers seem to have less influence than the leaders of corporations and professional societies, who do much more lobbying from one sector to another. There are some exceptions, such as the editors of the *New York Times* or the *Washington Post*, who can by persistence keep a topic on the agenda. But many trial balloons in the science columns fall flat.

Wolpert: It requires quite a high degree of conceptualization to say what scientific study means. Have you also asked people to compare two studies and say which they think is scientific? They might do much better with that kind of assessment.

Miller: We are continuing to explore the method. We have continued to use this measure because it gives us a good base point for comparisons, but we are also exploring other measures.

Laetsch: When Piaget was the godfather of science education in the United States, science educators measured whether young people could handle abstract concepts, and they made all sorts of definitions and classifications based on their supposed ability. It now appears that a lot more young people can deal with abstract concepts more readily than was previously assumed. One wonders, then, about why the numbers of those who are scientifically literate appear not to have increased even though most people have had information in the formal sense.

Presumably there is a failure not in information but in communication. That raises all kinds of issues, one of which is the definition of literacy. Is it important, for example, for people to understand that DNA is an aspect of the hereditary mechanism? Or is it more important for them to understand something about levels in the food chain and so on? If you start out teaching about DNA to the uninitiated you may well have some difficulties in communication. What does this tell us about the whole communication mechanism?

Miller: We have to be careful about assuming that people have had a particular kind of experience. Our high schools have adopted a smorgasbord approach to their curricula—students can choose their own courses. Last year only 15% of high-school graduates took a physics course and only 30% had a chemistry course. Those figures are lower now than they were 20 years ago.

Laetsch: But most states are now building science courses up again. In California secondary-school students must now have two years of science. Before 1957, though, many high-school students had three years of science.

Bodmer: How many learn mathematics?

Miller: Only about half of our high-school graduates have had a year of algebra. The number doing calculus or advanced mathematics is under 10%. The smorgasbord approach has had a much worse effect on mathematics than on science. The irony is that many mathematics teachers are against making algebra obligatory; they don't want to teach the 50% who don't choose to study it.

Tisdell: I agree that in the 21st century science and technology are likely to be more important than they are now. The problem is that we now live in a more dependent world and a lot of other things such as defence are important too. Do the attentive individuals in different areas belong to similar groups? What implications does this have? How do we get attentive individuals into a particular group? Are we going to train one group to keep an eye on science and technology policy, another group to keep an eye on defence, another group to keep an eye on economics, and so on? It is quite impossible for any of us to be well informed on everything.

Miller: Attentiveness as a process is very pluralistic. Most people are attentive to only one or two areas, depending on their occupational or academic interests. The college experience defines this much more than the high-school experience does. Being scientifically literate, however, has implications throughout foreign policy or economic policy or agricultural policy and so on: a knowledge of science enables people to follow those issues better.

Tisdell: People may not be attentive to an issue but if they become interested in it they can read up on it and become involved in it. Attentiveness is not a static matter.

Reference

Miller JD The American people and science policy. Pergamon, New York

Science and scientists: public perception and attitudes

GOÉRY DELACÔTE

CNRS, 15 Quai Anatole France, 75700 Paris, France

Abstract. Results from a recent questionnaire sent to about 100 000 French schoolchildren indicate how science and scientists are perceived by the age group 10–15 years. The answers from 3000 children show that they (1) are highly motivated to answer a questionnaire about science; (2) regard scientists as competent professionals; (3) attribute the most important scientific discoveries to the field of medicine (25% of replies); and (4) think that television is the best medium for disseminating scientific information (rather than school, for instance).

1987 Communicating science to the public. Wiley, Chichester (Ciba Foundation Conference) p 41–48

On 15 November 1985 *Okapi*, a monthly journal for readers aged 10–15 years, published a questionnaire entitled 'Science, what is it to you?' in collaboration with the Centre National de Recherche Scientifique (CNRS). *Okapi*, published by Bayard Press, has a circulation of 100 000 copies, and over 6000 questionnaires were returned to the editorial staff. The results given here are based on analysis of the 3060 questionnaires that were returned before 31 December 1985. The questionnaire was aimed at readers aged 10–15 years but the sample consisted mainly of those aged 12–13. More than half the questions were 'open' questions (Table 1), which allowed these young readers a good deal of freedom to express their opinions.

Young readers, highly motivated

Science is clearly a major preoccupation for this group. The large quantity of letters and additional explanations they sent in provided evidence of their interest in all aspects of scientific research. Although nearly all of these youngsters are fascinated by science, they nevertheless adopted a very realistic attitude towards it, as their answers show.

TABLE 1 **Translation of the questionnaire answered by (mainly) 12–13-year-old children**

---

(1) *How do you feel about progress in science?*

confident	a lot	a little	not really	not at all
afraid	a lot	a little	not really	not at all
fascinated	a lot	a little	not really	not at all
indifferent	a lot	a little	not really	not at all
enthusiastic	a lot	a little	not really	not at all
Other answers				

(2) *List the three most important discoveries, in your opinion:*

(3) *Rank the following countries from 1 to 8, according to which you think are the best in science:*

China	United States
France	England
Soviet Union	Germany
Australia	Japan

(4) *Here are four adventures: if you were invited to join in one of them, which one would you choose?*

Exploring the ocean floor
Travelling into outer space
Communicating by telepathy
Going back in time
Why?

(5) *What is a scientific researcher to you?*

A dreamer	or	A man of action
A loner	or	Someone who works as part of a team
A benefactor of mankind	or	An egotist working for his own pleasure
Someone with a lot of ideas	or	Someone who studies reality in detail

(6) *Do you think science will succeed in making man immortal?*

Yes
No
Why?

(7) *From where do you find out the most about scientific discoveries?*

Television
Specialized magazines
Comic books
School
Books
Museums, exhibits
Discussions with your parents, or with your friends

(8) *To succeed in your career, which subject do you think is better to be good in?*

History   Physics
French   Computer science
Mathematics   Languages
Sports   Biology

(9) *Which would you like to know the most?*

If there is life on other planets?
How man was created?
How the future will be?
Why?

URGENT, DISCOVERY

(10) *What, in your opinion, is the most urgent discovery science should make now?*
Why?

(11) *What do you think science will have succeeded in doing 20 years from now?*

Curing cancer?   yes/no
Building cities in outer space?   yes/no
Blowing up the earth?   yes/no
Eliminating hunger in the poor countries?   yes/no
Making daily life easier?   yes/no

(12) *Would you, personally, like to do scientific research later on?*
Why?

First name:
Age:
Sex:
Grade:
Number of brothers and sisters:
Father's profession:
Mother's profession:
City:
Postal code:

The scientist: a competent professional

In agreement with the largely positive images of science reflected by their answers, the children felt that a career in research leaves little place for being irresponsible, eccentric, or a pure genius. Far from scientists being regarded as 'absent-minded professors', they are considered to be competent, rigorous, precise. And they are no longer loners but work as part of a team. They are thought of as men of action more often than dreamers (48% versus 16%) and as benefactors to all humanity (64%), not egotists working for their own pleasure (3%). In addition, 'moral' and 'humanitarian' connotations were often associated with scientific research.

Medicine, veritable showcase of scientific progress

The scientific discoveries spontaneously mentioned as the most important give some idea of the image these children have of science, as do their priorities for action. The three fields most often mentioned were medicine (26% of the sample), the conquest of outer space (15%), and the use of natural resources and alternative sources of energy (13%). Medical discoveries are regarded as an almost obligatory prerequisite for further scientific progress. The answers concerning discoveries already made generally focused on the importance of vaccines and antibiotics. As for new discoveries, an overwhelming majority favours the fight against cancer and AIDS. Almost one in two youngsters explicitly mentioned the names of these two diseases as being the most urgent to control. This impressive consensus of opinion probably comes from a very large exposure to news reports on AIDS in the media, but it also comes from the shock of personal experiences involving cancer in their immediate families.

Although the children are greatly affected by disease, they do not expect medicine to make human beings immortal (82% think that is neither possible nor desirable). They do, however, expect medicine to improve health, to increase life expectancy, and even to solve certain social problems such as famine, war or injustice. They regard a 'serum against violence' or a 'vaccine against racism' as within the realms of possibility.

Television, the favourite medium for scientific information

Seven out of ten youngsters stated that most of their scientific knowledge comes from watching television. Other sources of information come far behind: specialized magazines (39%), books (30%), museums, (26%). Only 25% mention school, but comic books and radio are hardly mentioned at all (5%). The predominant role of television tends to increase even more with age (56% at ages 9–10 years, 79% at age 15 or over), while the interest in books declines.

It may seem strange to speak of the 'TV effect', as so few progammes deal with science on TV, but to scientific programmes as such (e.g. 'Animal Life', 'Future of the Future' or medical programmes) must be added the images from news broadcasts on current events, whether on earthquakes, the explosion of the space shuttle Challenger, or frozen embryos. These young TV viewers state without ambiguity that events take on more intensity and more realism when seen on television than on any other media, especially when an event is spectacular, even tragic. 'At school our history teacher told us about an earthquake in Mexico. I think it's better on TV because they show us the pictures'.

Acknowledgements

I would like to thank Anne Muxel-Douaire and F. Tristani for carrying out this investigation on behalf of CNRS. Further information may be obtained from them at CNRS headquarters.

DISCUSSION

Laetsch: What aspects of science interest adults in France?

Delacôte: I suspect that the same aspects interest them as interest children.

Laetsch: People working in science museums would probably agree with that. Although many people think we have to communicate differently to children and adults, there is the basis for a similar kind of communication.

Delacôte: The adult population in France is very interested in everything to do with medical research.

Bell: Given the large proportion of girls in your sample, do you think that the positive view of science among French children reflects a gender or a cultural characteristic? In New Zealand, because of our environmental and nuclear policies, science is regarded as being destructive and girls tend to be more interested in the human and social issues associated with science than boys, on the whole.

Delacôte: The main differences between the answers are more closely correlated with age than sex. It would be interesting to see whether their views alter much with age.

We are often asked what the benefits of research are and it is always difficult to explain that it is an ongoing process. People look for products, such as vaccines and so on, which they regard as a kind of magic. Yet the process itself is as magic as the products.

Wolpert: The answers you got show that the children were not very interested in science in the sense of acquiring more knowledge. They were enthusiastic about technology, that is about the product. We should not confuse science and technology.

Delacôte: Exactly. The questions were about science but the answers were about technology.

Bodmer: Why do you think science and technology should be separated?

Wolpert: To me, science is about knowledge and technology is about invention and its application. Historically the industrial revolution had little relationship to science. I would argue that we should not confuse knowledge with its applications.

Delacôte: The relationship between science and technology is indicated by the public's expectations of science.

Wolpert: One of the most important tasks of science communication is to make that distinction clear.

Caravita: You hinted that this confusion between science and its benefits is irrational, but it seems to me coherent and in line with the views that science puts out about itself.

Hearn: It is not surprising that at the age of 10 to 15 years schoolchildren identify with the end-point and only later appreciate the process leading up to the end-point. I was struck by the enthusiasm and optimism of some of the students who answered this questionnaire. Do you feel that they saw a personal future for themselves in science?

Delacôte: No, but they chose CNRS as a place to work, in preference to IBM, Havas and other commercial organizations. In France CNRS is perceived as representing all science.

Hearn: How important is that unified image of science?

Delacôte: I don't think it plays an important role in the choice of work. The problem lies rather in the view teenagers have of large bureaucratic organizations.

The conflict between process and product is as true in the domain of science as it is in technology.

Dixon: Are there any earlier surveys with which you can compare your findings?

Delacôte: The survey I described was the first to use such a large sample.

Evered: International comparisons would be interesting. The school system in France is somewhat different from that in most other European countries. Also, popular science magazines have a much higher circulation in France than in other countries in Europe.

Delacôte: International comparisons would certainly be useful, and a follow-up in France would be useful too.

Bodmer: If that is done I hope you will use a different sampling basis. It is important to have a proper sample of different educational levels and so on.

Delacôte: We don't claim that this is a piece of research but a piece of information.

Laetsch: We did a longitudinal study some years ago on how children perceive foods and their quality and whether their views change with age (from age 6 to 16 years) (Breazeale 1984). All the children had had some formal education in nutrition. It turned out that their perceptions did not change from the time they were young up to the time they were teenagers. The pattern of misconceptions closely followed the pattern of Saturday morning television, when a lot of nutritional advertising is done during programmes for children. The advertisers provide a mish-mash of nutritional information and the children pick this up rather than what they are supposed to learn in school. In the

USA people take in a lot from advertisements—which is why everyone coming out of the high schools now wants to be a computer scientist. The same effect may be at work in France.

Whitley: Were the questions in your survey based on recall?

Laetsch: No. The children were shown various foods and asked to put them in different categories. Their responses didn't make sense until we realized the connection with television advertising.

Delacôte: We have no detailed information on which television programmes might have influenced the children in our samples. I agree that advertising is an important way in which children obtain information, rather than through magazines.

Friedman: Studies in the United States by the National Assessment of Educational Progress show a decline in interest in science the longer the children remain in school.

Delacôte: In our sample it seems that less information is obtained in school as the children grow older.

Friedman: In France they still seem to want the information but in the USA they have less interest in learning science at the later age.

Falk: It depends on what they consider to be science.

Miller: It is not entirely clear from the way the question is written in the NAEP studies whether students have less interest in science or less interest in science courses. Given the smorgasbord of courses in the USA, I think we inadvertently give students the impression that science is only for the very best and brightest and this puts them off science.

Lucas: Even within those courses the interest declines over the time they are taking the courses. It is a common finding that interest declines from the age of 16 onwards, probably in any subject (for examples, see Lucas & Broadhurst 1972).

Miller: That is not true of college students. The more science they take, the more interested they get.

We are doing a longitudinal study for the National Science Foundation of 7th and 10th grade children, asking mainly about their interest in science and mathematics but including a whole range of life choices. Parents, teachers and counsellors are being questioned, as well as the students. The gender difference is perceptible by grade 9 and very pronounced by grade 12.

Laetsch: Interest in science often declines as children do more science in high school, yet the universal response to the crises we have is to provide more science, or more mathematics, or more chemistry, and so on.

Bodmer: The answer is not to teach more science but to teach it better.

Laetsch: Yes, but the emphasis is on quantity. The quality of the communication of what is provided is often more important.

Miller: There are many schools in which exposure to science increases the interest of students. But across the country, high-school science is taught in such a way that it has a negative effect. The objective should be to multiply the stimulating experiences, not to put more students into bad science courses.

References

Breazeale VD 1984 Development of concepts about food and nutrition. Ph.D. thesis, University of California, Berkeley

Lucas AM, Broadhurst NA 1972 Changes in some content-free skills, knowledge and attitudes during two terms of grade 12 biology instruction in 10 South Australian schools. Australian Science Teachers Journal 18(1):66–74

The role of schools in providing a background knowledge of science

BEVERLEY BELL

Curriculum Development Division, Department of Education, Wellington, New Zealand

Abstract. Two aspects of science education continue to be in sharp focus for science teachers. Firstly, children come to formal science lessons with their own understandings of their natural and technological worlds. Secondly, scientific knowledge is expanding at an explosive rate. The role of schools, therefore, is not to transmit knowledge but to encourage and facilitate students to make better sense of their biological, physical and technological worlds; to help students to be responsible for their own learning; and to promote a view of science as a human activity. Thus, in promoting the personal growth of science students, schools can help them to extend their own knowledge of science during and after their formal science education.

1987 Communicating science to the public. Wiley, Chichester (Ciba Foundation Conference) p 49–63

As a science educator, several thoughts come to mind as I consider the idea of a 'background knowledge of science'. These include the exponential rate of growth of scientific knowledge; the intuitive ideas students bring to a lesson and the powerful influence these ideas have on the learning outcomes; and the groups of students who see the scientific knowledge learnt in school as boring, irrelevant and not useful to their lives. It is these last two points I would like to elaborate on in this paper.

Changing one's ideas

Schools do not have to provide a background knowledge of science. Students bring to science lessons their own ideas about their biological, physical and technological worlds, for, like scientists, they have been trying to make sense of their world from an early age. However, their ideas tend not to be the current scientifically accepted ones and have been called 'children's science' (Osborne & Freyberg 1985), alternative ideas (Gilbert & Watts 1983) and children's ideas (Driver et al 1985). The role of schools is therefore not so much in giving or transmitting scientific knowledge but in helping students to change their ideas so as to make better sense of their world. This process of conceptual change involves basically two stages (Bell 1984). The first is that of *constructing* a new idea; this involves ·generating links between incoming stimuli and existing knowledge (Osborne & Wittrock 1983). If the existing knowledge is not scientifically acceptable, unintended ideas may be constructed. For example, a student who had only an everyday meaning of 'animal' (Bell 1981) and hence considered only the large, four-legged furry creatures as animals read the text:

The world of living things can be divided into two main groups: animals and plants

and commented:

Oh yeah, but some things wouldn't come under one of them. [What sort of things?] At the moment, a spider's not an animal, wouldn't come under plant, and a worm, and things—insects.   (Bell 1984)

In helping students to construct new ideas—for example, the scientific conception of 'animal', gravity or electric current—we need to be aware of the students' existing ideas and how they influence what understandings the students will construct. Telling students the new idea is not enough, for each student has to construct for herself or himself the new idea and in doing so will draw on existing knowledge, which may or may not be intuitive ideas.

The second part of conceptual change is *accepting* a new idea as part of one's belief system. Constructing a new idea does not automatically mean accepting it (Strike & Posner 1984). For example, a student read the text:

A spider is an animal as it eats flies for food

and commented:

I still think spiders are insects, not animals. Don't know why. I just don't feel a spider's an animal.   (Bell 1984)

The decisions to accept or reject a new construction is not just a rational decision. There may be a rational checking of plausibility and consistency to see if the construction logically makes sense, and to see if the environmental stimuli match the store of existing knowledge. But there may also be the affective aspect of a decision to accept or not accept the new construction as part of one's own knowledge. Does the new construction lie easily within the learner's existing knowledge in an intuitive way? Can the learner believe in it? (Strike & Posner 1984.)

Conceptual change can be a slow process, for despite up to five years of formal science education, many students may still hold their intuitive ideas. The ideas learnt in science lessons may be kept separate from their beliefs about the world and used only in the context of a science lesson or examination (Osborne et al 1983).

Claxton (1984a) asserts that resistance to conceptual change may be due to a 'personal stake in believing that things are a certain way, that is, if I *identified* with a point of view then I cannot give it up without also giving up something of myself. If I am identified with my own competence, for example, then even to admit I might be wrong may be too scary. Learning in this case becomes a *threat to myself*.

Claxton adds that learning may be a threat to a person's knowledge structure if it involves radical upheaval of everything the person knows. It may also be a threat to a person's social stability in that to change may put the person in conflict with society (including the church), family and friends.

Claxton (1984b) advocates that, as insecure feelings are an integral part of any learning situation, we may need to help students to develop a greater tolerance for insecurity, to accept that feeling unsure and anxious is an acceptable and expected part of being in a learning situation. This suggests the need for a supportive classroom atmosphere, in which different views are valued and given attention, and in which it is acceptable to express doubts and hesitations. The view of scientific knowledge that is often reinforced in science classrooms is one of 'absolute truth' and not one of a tentative nature. Many students (and teachers) perceive a major outcome of science lessons to be getting the 'right answer' or the 'correct result or instrument reading'. This view works against an open discussion of different conceptions and feelings in the process of learning. Likewise, some assessment strategies may not promote the kind of supportive atmosphere desired.

West & Pines (1983) assert that learners' feelings are an important component in conceptual change and 'are integral parts of what learning *is*, and not simply motivational, attitudinal or affective antecedents upon which learning depends'. They describe four groups of feelings: feelings of power that come with the ability to identify, predict and make sense of the world; feelings of satisfaction that come with patterning and ordering imposed on a complex world; positive feelings associated with aesthetics; and comfortable feelings of personal integrity and lack of dissonance.

In summary, helping students to change their ideas (towards scientific ideas) involves helping them not only to construct new ideas but also to accept them. As conceptual change can be threatening, we need to take into account the feelings associated with the learning.

Being responsible for one's learning

If scientific knowledge is increasing at an exponential rate, and all of public knowledge cannot become personal knowledge, then the role of schools is not only to help students to construct scientific understandings but also to help them to take charge of their own learning processes. Being responsible for one's learning encompasses several aspects:

(a) Conceptual change tends to be enhanced if learners are aware of the learning process itself. Baird (1984) argues that learners need to be helped to develop and use skills to monitor what they do and do not know, what they are learning and why, and what it means to be a learner—that the learning process is not a passive one but one which requires their time, energy and commitment to construct, evaluate, accept and subsume new ideas.

(b) Conceptual change may be promoted and the change may be less threatening if students are given more responsibility in choosing their own areas of investigation (Boomer 1982). This is not to suggest that students should necessarily be given all the responsibility but that the curriculum should be negotiated between student and teacher. A negotiated curriculum is a way of maximizing learning that is useful and relevant *for the student*.

(c) Learning also tends to be enhanced if students are helped to find answers to their own questions (Biddulph & Osborne 1984). This involves helping them not only to explore their world but also to ask useful questions and to use a variety of sources of information to obtain answers. The traditional role of the teacher or the 'expert' becomes that of facilitator and the role of the student changes from a passive to an active learner. The notion of a spiral curriculum takes on a new look if the spiral is determined by the nature and direction of the questions of students (not of the teachers).

In my experience, it is this student-paced and directed learning (often within the broad guidelines of the teacher) that fosters the sense of curiosity, inquiry, creativity, excitement, awe and wonder that we would wish to characterize science lessons.

Belonging to science

Many science students do not feel a part of science. In my country, girls (and Polynesian students) often feel excluded from, rather than included in, science, and they feel alienated from it. Girls typically tend to see science, especially the physical sciences, as boring, not useful, and detached from their world. In increasingly scientific and technological societies this is of concern for education, vocational and social reasons.

The reasons for this alienation have been outlined elsewhere (Bell 1987) and I wish here only to summarize them. Firstly, girls and boys bring different experiences, interests and concerns to a science lesson. Hence, the existing ideas which are used to construct new ideas are different. For example, the interests and experiences of girls often centre around pets and plants, whereas those of boys tend to be associated with weapons, machinery and space. Typically, it is the experiences of the boys that are used to illustrate scientific ideas and skills in a lesson, not the experiences of the girls. It is not surprising, then, if the girls feel their worlds are outside the concerns of science.

Secondly, girls and boys tend to choose to study science for different reasons (Harding 1985). While many boys are drawn to science because they want to understand phenomena and devices, girls tend to see science as offering solutions to human problems. Most school lessons do not present science in its social context and many girls (incorrectly) perceive science as useless and irrelevant to their concerns or social issues.

Thirdly, boys tend to dominate classroom talking time, use of equipment and the attention of teachers. Not only may girls feel undervalued but they may also dislike the competitive, not cooperative, atmosphere engendered.

Lastly, the nature of science and how it is portrayed in classrooms need our attention (Fox-Keller 1985).

To help all students to feel they belong to science, the experiences and concerns of all students, not just those of the boys, need to be given importance in the classroom.

And finally

Three general aims of eduation have been important to me as an educator (New Zealand Post Primary Teachers' Association 1969):

the urge to enquire;
concern for others;
the desire for self respect

For me, these can be translated in terms of science education as the following aims (Bell 1986):

(1) To help students make better sense of their biological, physical and technological worlds
(2) To encourage students to keep on asking questions about their world and to help them seek answers to those questions
(3) To help students to be responsible for their own learning
(4) To promote a view of science as a human activity

These aims are not all exclusive to science education, thus reflecting that science education is a part of the wider education of students, and that the personal growth of students is a part of science education. This is reinforced in the fourth aim: to promote a view of science as a human activity. This 'humanness' can be expressed in several ways (Bell 1986):

● Scientific understandings (models, interpretations, concept theories) are constructions produced by the human mind in an effort to make sense of phenomena and events.
● Science is an activity done by a person or usually by groups of people. Working cooperatively with other people is part of science, for it involves communication with others.
● Science influences the lives of people—for example, science is responsible for antibiotics and nuclear weapons.
● The ideas, beliefs, values and culture of scientists to varying degrees influences scientific activities and understandings—for example, current research on reproductive technology and weaponry.

Overall, the role of schools in providing a background knowledge of science is one of ensuring that students feel good about learning science and about science itself. It is how people *feel* about these that is a strong factor in whether they continue to investigate their world from a scientific perspective as adults. Feeling good about science learning and science itself includes:

● Students changing their ideas and behaviour in a non-threatening and supportive atmosphere
● Students feeling that the ideas, interests, concerns and questions which they (not the teacher) bring to the classroom are of value
● Students being encouraged to ask questions
● Students learning in a way that enhances self-esteem and empowers them to take control of and responsibility for their own learning
● Science being seen as having a concern for people and not just as associated with destructive forces, and
● Science being seen as something done by people—people who feel as well as think.

Acknowledgements

I wish to thank the following people, whose discussions helped to clarify for me many of the ideas in this paper: the Form 1–5 Science Revision Committee, Andy Begg, Colin Percy and Malcolm Carr.

References

Baird J 1984 Improving learning through enhanced meta-cognition. Unpublished D.Phil. thesis, Monash University, Melbourne, Australia
Bell BF 1981 When is an animal, not an animal. Journal of Biological Education 15(3):213–218
Bell BF 1984 The role of existing knowledge in reading comprehension and conceptual change in science education. Unpublished D.Phil. thesis, University of Waikato, Hamilton, New Zealand
Bell BF 1986 Science working in classrooms: the role of practical work. Proceedings of SCICON, New Zealand Science Teachers Association (mimeo)
Bell BF 1987 Girls and science. In: Middleton S (ed.) Women, education and schooling in Aotearoa. Otago University Press, Dunedin, New Zealand, in press
Biddulph F, Osborne RJ (eds) 1984 Making sense of our world: an interactive teaching approach. University of Waikato, Hamilton, New Zealand
Boomer G (ed) 1982 Negotiating the curriculum, Ashton Scholastic, Sydney
Claxton GL 1984a Teaching and acquiring scientific knowledge. In: Keen T, Pople M (eds) Kelly in the classroom. Cybersystems, Montreal
Claxton G 1984b Live and learn. Harper Row, London
Driver R, Guesne E, Tiberghien A 1985 Children's ideas in science. Open University Press, Milton Keynes
Fox-Keller E 1985 Reflections on gender and science. Yale University Press, New Haven, CN
Gilbert JK, Watts DM 1983 Concepts, misconceptions and alternative conceptions: changing perspectives in science education. Studies in Science Education 10:61–98

Harding J 1985 The making of a scientist? Centre for Science and Maths Education, Chelsea College, University of London

New Zealand Post Primary Teachers' Association 1969 Education in change. Longman Paul, Auckland, New Zealand

Osborne RJ, Freyberg P 1985 Learning in science: the implications of children's science. Heinemann, Auckland, New Zealand

Osborne RJ, Wittrock MC 1983 Learning science: a generative process. Science Education 67(4):489–508

Osborne RJ, Bell BF, Gilbert JK 1983 Science teaching and children's views of the world. European Journal of Science Education 5(1):1–14

Strike K, Posner GJ 1984 A conceptual change view of learning and understanding. In: Pines AL, West LT (eds) Cognitive structure and conceptual change. Academic Press, New York

West LT, Pines AL 1983 How 'rational' is rationality? Science Education 67(1):37–40

DISCUSSION

Screven: The motivation and learning of museum visitors reflect many of the things you listed; in museums these may lead to motivational problems and major distortions of exhibit messages. Learners in science museums are voluntary and operate on their own time and on their own terms. Given this, museum experiences must be 'fun' experiences. The 'fun' in museums is based on visual and participatory exploration, social activities, curiosity, surprise, challenge, and self-directed learning. But these can often be obtained in science museums without the visitor learning anything important—that is without attending to the concepts the exhibits are intended to convey. Fun and excitement are often observed at science exhibits, but these do not necessarily mean that science education is taking place. At many science exhibits, there are things to push, turn or manipulate that produce motion, lights or actions that most people enjoy, but many of these exhibits are not designed so that these depend on noticing certain things, discovering generalizations, and so on. The teaching objectives of exhibits should be linked with the intrinsic fun of interacting with them. Instead, the fun usually can be had with or without learning anything very important.

But planners who worked so hard to produce these exhibits find it hard to imagine that this content will not be understood when visitors interact with them. Roger Miles (1986) has described the perception of the museum audience by curators and scientists as being very different from the actual audience. Poor understanding of museum audiences results in a lot of otherwise good exhibit content being ignored or poorly understood and in a lot of misdirected visitor activities. Exhibit planners must become better acquainted with their actual audiences, what motivates their cooperation, how they behave, perceive and use science exhibits, their social and voluntary nature, their visual and participatory bias, and the misconceptions that affect the ways they relate to and perceive exhibit content (Screven 1986). Some of this information is available (Screven 1984) but is seldom utilized during planning.

Falk: Your depiction of what is required in formal education stands in stark contrast to most of the requirements I have heard portrayed, Dr Bell. How do you square what you are suggesting with the current realities?

Bell: The research showing the kinds of ideas students learn or do not learn in the lessons is a very powerful influence in helping science teachers in New Zealand to reflect on what is happening in their classrooms. When I worked in Leeds (with the Childrens Learning in Science Project) we found the same thing. What we thought students were getting out of a lesson was not what they actually were getting out of it. If only a third are actually learning scientific concepts, we have to stop and examine why. Once the need for change is accepted, there is a way forward.

One of the issues we then started to expand on is the inductionist view that if you have the data—the scientific evidence—you can extract the knowledge from it and come up with the scientific generalization. That is not what happens. Instead, students bring their own ideas into the classroom and those ideas are far more influential in what meanings they construct. Unless we address those alternative ideas, students will tend not to change them.

Falk: In the USA, the expectations of the educational establishment appear to be diametrically opposed to your model, both in terms of how they present instruction in schools and how they determine success.

Bell: In New Zealand things are moving slowly towards this. We have to draw a parallel between conceptual change in students and conceptual change in teachers. The teachers need opportunities for in-service and professional development that allows them to change their ideas about their role in the classroom in a supportive atmosphere.

Macdonald-Ross: A communication may trigger some underlying misunderstanding that may have quite complex causes. To change the ideas of a person you have to unravel what lay behind the confusion. The process of analysing what is causing this misconception is part of the art of communication and is a prerequisite for real change in an individual. The misconceptions may have quite deep roots and you have to strike at those first.

Tisdell: In economics the amount of formal knowledge has increased greatly. If we try to get through all this, students have very little time left to read for their degree, apart from reading textbooks. Basically, many students are just interested in doing the formal work required to get through the final magical examination that determines their entry to university. Many teachers are also like that, because the system is nice and structured. I like your approach but how can we blend these two things together when we have state-wide examinations?

Lucas: You change the examination system to reward those values. It is easier to do that with state-wide examinations than otherwise. You have a very strong lever.

Tisdell: The system in Australia now allows for assessment by teachers but there are some problems with that.

Lucas: Beverley Bell is proposing a different set of goals for science education, concentrating on the processes people use when they are learning science. The notion of responsibility for one's own learning is a good link with the communication of science to the public as a whole. Can people be helped to learn how to learn and how to take responsibility for their learning outside the school context? That is what we are interested in really, continuing to learn after those compulsory years of school.

Bell: My ideas relate to in-school learning, but are applicable to informal learning. The process of learning is the same for the students.

Tan: It is true that teachers cannot run away from the examination. They have to finish the syllabus and make sure that the principal, the parents and the ministry are kept happy. Research in science education has pointed out that cooperative activities have favourable effects on science learning. Lam-Kan (1985) did an experimental investigation of this, working with a group of 180 secondary school students, all girls, aged about 13 years, which is just before they are streamed into science or arts. This teacher assigned the girls, who were not particularly interested in science, to five groups: two control groups, two experimental groups which had enrichment lessons, and a Hawthorne control group. On excursions the Hawthorne group went only to places of interest that were totally unrelated to their science lessons. One experimental group, one control group and the Hawthorne control group were pre-tested and all five groups were post-tested on school examinations. The two experimental groups participated in the enrichment activities available at the Singapore Science Centre. The control groups had no out-of-school lessons. The activities at the Centre were designed to complement topics taught in the school at the time of the experiment during the final term the students spent at school. The study depended largely on the school science teacher and the Science Centre officers who collaborated in planning the course.

The pre-tests and post-tests used two standardized instruments, the Reed Science Activity Inventory (Cooley & Reed 1961), consisting of an inventory of 59 science activity items, which is used to give a measure of general science interest, and the Cooperative Science Test (COST) (Educational Testing Service, Princeton 1963) consisting of objective multiple choice tests used to assess attainment of science concepts. Science learning improved significantly ($P < 0.01$) in the experimental groups, and interest in science also increased significantly ($P < 0.01$). But a lot of the girls commented that school science was very boring and too examination-oriented. Something has to be done about the examination systems.

Evered: Did the questions in the two tests relate to facts rather than perceptions?

Tan: They related to both.

Laetsch: One might argue that people shouldn't have science during the regular school years but should go on to continuing education.

Thomas: In my world of informal adult education many people are there purely for enrichment courses. Adult students who come to such classes are there voluntarily and they have the option of negotiating their own curricula. Once a group has formed it can determine for itself the direction of its discussions; these are a crucial part of the exercise in adult education. Tutors are there not so much to teach as to enable students to learn, through answering questions, entering into discussion, and so on. The best tutors ensure that students feel positive about the process by showing them that their interests and ideas are valuable. In the sciences that is not always easy to do, because a lot of the questions and comments may be 'wrong', in a scientific sense. This sort of science teaching is time-consuming and can only be done effectively with small groups; that is a big snag in times of limited resources.

Even if the ideas that come out in discussion are correct, they may still have to be handled delicately. I once had a student who discovered a variant of one of Kepler's Laws. I had to break it to him gently that this had already been done—a slightly different teaching problem from most. Usually one has to deal with people's ideas very gently, particularly when dealing with adult students.

Laetsch: You can use that as an example showing that in science we are irrelevant individually, because discoveries will be made by someone, whoever it may be.

Nelkin: A whole genre of post-school science education goes on in areas such as the extension services, where people have economic stakes. Successful farmers are continually learning science and scientific techniques, and the same is true in industry and health. When something is wrong with you and you try to become informed, you can learn a lot of science in the process.

Lucas: How do you prepare your captive audience in schools to take part in areas in which they may have a stake?

Nelkin: I worry that too little attention is given to teaching the process of science. A famous article is called 'Should the history of science be rated X?' (Brush 1974). The notion is that if scientists themselves think too much about the process of science it would paralyse them because the intuitive process which is fundamental for successful science would break down if they analysed the process too carefully. The best the education system can do is to get people to be not afraid, not in awe, and to remove the notion, perpetuated by the press and the school system, that science is something arcane.

Gregory: How literally do you want to take the issue of schools not transmitting knowledge? How far are they modifying or refining existing beliefs? When one first learns about lines in the spectrum a whole new world opens up and one realizes that the chemistry of the stars is related to the bunsen burner flame or to the effect of salt on a candle flame. This comes not so much from a book as from doing it. In the arts you are supposed to find out about sexuality, affection, aggression and so on from Shakespeare and others, and this refines you. Science is not like that at all. You are in a completely new world that you hadn't dreamt of.

If you ask young children what holds the moon up they think you are mad. They haven't thought of it as a question. They have no views about what holds the moon up or about cosmology and they start from a total vacuum. Science is very different from the arts when it presents worlds of wonder and questions one is not taught about at all as a child until they come up in a science class, preferably a practical class.

Bell: I don't think scientific ideas are completely new to students. If they were, students couldn't make any links and therefore start to understand it. Something in your existing knowledge must enable you to make links between what you see or hear and what you already know, if you are to make some sense of it.

In Leeds I was talking to 15-year-old students about plant nutrition. Some had done biology and some had not. In response to one question, they often talked about the ready-made food being sucked up through the roots. They called the water and minerals food and they were quite happy that plants got food and so on from the soil. They talked about food for plants just as we might talk about food for humans. In response to another question they said that plants make their food during photosynthesis and they gave the chemical equations for what was happening. Then I challenged them about their response to the first question, when they said plants sucked food up through the roots. They said 'Ah yes but plants do both. When the sun is out photosynthesis is going on and when it is night food is sucked up through the roots.' They didn't change their original ideas but added something to them. They were able to link the new ideas to what they already knew without changing their existing ideas substantially.

Caravita: The integration between in-school and out-of-school education can promote conceptual change. The two contexts have their own characteristics in terms of motivational power, interactional possibilities, cognitive skills implied, time-schedule, etc. Both should contribute to enlarging the range of phenomenological evidences and exemplars that people can take into account in building up their mental models about reality, which have to be powerful tools for generalization, prediction, explanation. Moreover people should be given the opportunity to confront their meanings and go beyond their idiosyncratic and subjective perspectives through negotiation and sharing in a social context where the views of science can play their role. Even the exposure to different systems for representing knowledge pushes towards conceptual change. This has to be taken into account in setting up displays in museums but also in schools, where teachers might discuss with the students the relationship between the symbolic code chosen and the characteristics of the object that has to be represented. When we pass from one representational system (e.g. a graphic one) to another (a numerical one) we are forced to analyse different features and their interrelationship within an object of knowledge: this produces abstraction.

Out-of-school education may afford chances for all these types of reasoning procedures and operations; then if teachers can and are able to make more use of it they would find a variety of resources at their disposal for further elaboration within the class.

Deehan: I too must take issue with the notion that schools are not there to impart knowledge. People who try to make expert systems that will allow machines to operate in the same way as a human expert are finding that they need to use enormous amounts of knowledge as well as applying logic or intelligence. If you put in these quantities of knowledge you find that the rules that the machine needs to extract meaning from the world are very simple. It is possible to go too far in the wrong direction and say that we have to give our children some means of acting like Aristotle or some other logical model, ignoring the fact that before you can apply these forms of decision-making you need knowledge. And to make sense of science you need an enormous amount of knowledge.

Bell: That was my point about existing knowledge. Without some related knowledge, it is difficult to make the necessary links. In this way, the alternative ideas get used just as much as the ideas of science. It is not a case of transmitting knowledge but of facilitating students to construct the knowledge for themselves.

Miles: Piaget uncovered some quite extraordinary explanations from children as to what makes clouds move and so on. The problem for those of us who work in museums is that if you simply present phenomena, people may provide their own explanations and see evidence in what you present for their alternative conceptions. It is quite common in museums to come across adults explaining the significance of something to children in terms that would astound the curators who designed the exhibits. Alternative conceptions are a real difficulty in the whole field of science education. I don't know if you have ideas from formal education that would help us in museums, or whether our problem is of a different order?

Caravita: You cannot expect people to dismiss their existing conceptions verified in everyday life immediately you expose them to new ideas.

Falk: That is exactly the point—people's conceptions are adaptive. The burden is on science educators to prove that their view of the world is better and more functional, more adaptive, than that of other individuals.

Caravita: But that is not always the case.

Macdonald-Ross: The links between prior knowledge and the knowledge that is being presented can be of many different kinds. If your existing idea is designated X and the new idea presented by a teacher is X', the simplest kind of relationship is one of consistency. If consistency is involved, a large number of X's might be tagged on to the X. On the other hand the link might be a generative one: if you are holding X^y you can predict that X' ought to be the case. If, instead of the expected X', you want to teach X'', you are in a more complex position. You need to show the learner that his or her preconceptions need to be changed. When you try and do that, you run across the real problems of teaching and learning. By contrast, when you are dealing with consistency you are dealing with prior knowledge which for all practical purposes can be disregarded. Although it is there and the links are many, they are not constraining. Therefore, in a sense, the learning can be treated as being new in the sense that Richard Gregory was talking about earlier.

Wolpert: To what extent is teaching authority-based? One could probably teach false science very easily. I am really frightened at the idea that most of what people know about science is what they learnt by rote. My students will learn almost anything just to get through examinations.

Lucas: We take people like that for teacher-training after university science departments have had a go at them. They can answer the expected questions but if we ask them to look at something in a slightly different perspective, they find it very difficult.

Tisdell: That is the difference between honour students and pass students.

Laetsch: We expect the person who is scientifically literate to understand all sorts of things. Maths majors will come and take a biology course but if you ask them to expand the binomial theorem they get blinkers over their eyes. Or a biochemistry major won't understand that a plant that is producing carbon dioxide is losing weight. How do we communicate between the little boxes in which we have all learnt our science?

Gregory: In Bristol I give courses for medical students on the history and philosophy of science and scientific method. I haven't found one student yet who knows the difference between induction and deduction. With all their A level grades they have a very poor idea about things like statistical significance and scientific method, or what it is to evaluate data. This is surely a tremendous criticism of school science.

Wolpert: Most undergraduates seem to believe that when you do an experiment you should have no idea of what result to expect, because that would prejudice the experiment. That is exactly the opposite to the reality of research.

Caravita: Many of the experiments run in schools are demonstrations, not experiments.

Wolpert: I always feel school teaching is better than university teaching.

Miller: Your idea that children bring models in their heads is very true. School children personify matters at very young ages: if you ask them how countries get along, they tell you how people get along. If you ask them how a poor country can get rich, they tell you that it is by getting a second job or something. On the other hand, there is a developmental stage when they become able to handle abstractions. Will the models change as the children develop the ability to process abstraction? Does the negotiation process you mentioned take into account the developing ability of students to see things in more abstract terms?

Bell: The issue of why some students in classrooms change their ideas and others do not is at the cutting edge of the research at the moment. I feel it is those students who are seeking answers to *their own* questions or problems who are most willing to change their ideas, rather than something to do with different kinds of thinking. For example, I recall a long conversation with a student who was confused about whether human beings are animals or not. She had held the alternative view that people aren't animals because they haven't got four legs, but she was beginning to challenge her own ideas. It was something she wanted to address and in the end she changed her ideas towards the scientific view.

Hearn: To what extent are we deficient in instructing and infusing scientific method and logic in schools in order to solve problems, induce confidence and lose the fear of science? If only 5% of the population is scientifically literate even at graduate level, it gives an impression that science is very difficult. Perhaps students do not appreciate that the tools are there to break down problems and to solve them in a logical way.

Bell: Wynne Harlen (1986) has addressed the issue of how we can help students to test their alternative ideas in a scientific way. She gives the example of the very commonly held view that electricity travels faster in a straight wire than in one that is curled, say, in household appliances. If you ask people how to test this they often say 'You just pull the wire straight' and see the difference. There is often no sense of comparison in the testing. The idea of comparison and fine measurements is part of science as I see it. Maybe we should be helping students to move from where they are at by looking at the process of testing out new ideas.

Hearn: When I lecture at sixth-form colleges, the teachers often say 'Don't make it difficult, because science is very difficult and they are not all that clever'. Yet the questions the students ask are excellent. Both teachers and students assume that science is difficult and that only the best students can go for it. I have seen this in cohorts of schoolchildren where the teacher has said 'If you want to do three sciences that is far too ambitious for you or for your class'.

I always start first-year lectures in university by emphasizing just how little we know and how much is left to be discovered. The students wake up when they hear this. Many of them seem to feel, when they leave school, that it is all in the textbooks with nothing left for them to do. If the excitement of discovery could be got over at an earlier level, it might reduce the fear and also encourage participation and enthusiasm.

References

Brush SG 1974 Should the history of science be rated 'X'? Science 183:1164–1172

Cooley WW, Reed HB Jr 1961 The measurement of science interests: an operational and multidimensional approach. Science Education 45(4):320–326

Harlen W 1986 Creativity and rationality in learning and teaching science. Inaugural Lecture, University of Liverpool

Lam-Kan KS 1985 The contributions of enrichment activities towards science interest and science achievement. Master of Education thesis, National University of Singapore

Miles RS 1986 Museum audiences. International Journal of Museum Management and Curatorship 5:73–80

Screven CG 1984 Educational evaluation and research in museums and public exhibits: a bibliography. Curator 27:147–165

Screven CG 1986 Exhibitions and information centers: principles and approaches. Curator 29:109–137

Interactions between formal and informal sources of learning science

A.M. LUCAS

Centre for Educational Studies, Kings College London, University of London, Chelsea Campus, 552 Kings Road, London SW10 0UA, UK

Abstract. People learn science from many sources: from schools and other formal institutions, from accidental encounters with scientific ideas, and from deliberate encounters with specially provided facilities such as museums. But none of these sources, formal or informal, exist alone. A schoolchild is exposed to the science in the media; the adult museum visitor may have some ideas about the topic already.

There is incidental evidence that people with better science backgrounds are more likely to attend science museums, read science news in the press and regularly watch television science programmes, other than natural history programmes. Although the data available are incomplete, difficult to interpret and to generalize, they do remind us that the interactions between sources of knowledge are complex, and that it is rash to make general assumptions about the knowledge that our pupils, our readers, our visitors and our viewers may already have or may gain from our efforts.

1987 Communicating science to the public. Wiley, Chichester (Ciba Foundation Conference) p 64–80

Some years ago one of my student teachers complained that 'The boys have seen so many reactions on television, with lots of sparks and explosions and dramatic changes that the few bubbles that come off calcium in water are very small beer indeed. How can I compete with their expectations of spectacular effects?' (Lucas 1981.)

At least twice I have overheard conversations between children and adults, presumably parents, as they left the Insect House in the London Zoo. The gist of the conversations was that the child's teacher had been wrong. When questioned by the adult, the child complained that the teacher said that insects have six legs, but that there were things in there that had many legs. (The house is a general invertebrate house, and displays spiders and millepedes in addition to insects.) When the parent suggested that perhaps the teacher was right, the child insisted that the Zoo said it was an *insect* house, and the Zoo must be right.

Both these cameos concern an interaction between formal and informal sources of scientific knowledge. In the first, there is a potentially inhibiting effect of informally presented science on the acquisition of knowledge in school; in the second, the school knowledge is rejected in favour of an (incorrect) interpretation of the message of an informal source perceived as authoritative by the visitor. There is no reason to believe that such effects are restricted to children. Indeed, one of the justifications given for strengthening school science is to give adults a base to help them continue to learn throughout their life (for an indirect version of this argument see Bodmer et al 1985; for a more direct version see Booth 1979).

In this paper I begin by outlining the nature of possible interactions between informal sources of science and formal schooling and then consider what evidence there is about these possibilities. I conclude by making some suggestions for research.

Possible interaction patterns

In order to discuss interactions between different sources of knowledge it is useful to categorize possible types of interaction, so that we can see the dimensions of the problem we are discussing.

I have previously argued (Lucas 1981) that interactions could be *facilitating*, with one source enhancing the (possibility of) learning from the other, *inhibitory*, with one source diminishing the effect of the other, or *neutral* in their effect. We can ignore the neutral interactions, for these will have no effect. The following possibilities remain:

Formal sources facilitate learning from the informal source
The informal source facilitates learning from formal sources
Formal sources inhibit learning from the informal source
The informal source inhibits learning from formal sources

It is also possible that the effects will be mutual. This means we need also to consider these additional possibilities:

Formal and informal sources are mutually facilitatory
Formal and informal sources are mutually inhibitory
Formal sources facilitate learning from the informal source, but the informal inhibits learning from the formal
Informal sources facilitate learning from the formal, but the formal inhibits learning from the informal

I am not confident that I can imagine convincing examples of the last two possibilities, but I can imagine simple possible examples of the other six. Some examples are given below; others can be found in Lucas (1981).

Formal facilitating learning from the informal
Principles of biological classification taught in school may facilitate learning the classification of plants from the British Natural History gallery in the British Museum (Natural History).

Knowledge about simple genetic ratios learnt in school may facilitate learning, from a blood grouping sequence in a detective story, the principle of exclusion of possibilities.

Informal facilitating the formal

On a family visit to the Zoo, close observation of monkeys feeding on a variety of food may enhance understanding of a later lesson on human tooth structure and function.

Arranging buoyancy bags in a dinghy according to the instructions and explanations in a yachting magazine may later facilitate learning about resultant forces in physics classes.

Informal inhibits learning from the formal

Incomprehensible labelling of the steam engines preserved in the Science Museum in London may so confuse young visitors that their teacher is later unable to motivate them to learn the operating principles of external combustion engines.

The everyday language use of 'mixed blood', 'blood lines' and similar terminology may interfere with pupils understanding gene theory in biology.

These examples also illustrate another categorization that is useful: informal sources of learning may be *deliberately* educative (the first of each pair), or *accidental*, sources from which it is possible to learn although they have not been intentionally created with this in mind.

It is much easier to study deliberate informal sources, such as the press, documentary television, and science museums. There is, however, an increasing literature concerning the ideas that children bring to school with them. Much of the literature on physical science topics is reviewed in Driver et al (1985), and Bell (1985) reviews some of the biological studies. These studies can be interpreted as indicating that the accidental sources of information can build up such strong 'alternative frameworks' of ideas that it may be difficult to replace them with the accepted scientific framework. For the remainder of this essay I shall be concerned almost exclusively with the deliberate informal sources of education, but these are not the only ones from which we learn.

Does school science facilitate use of informal sources?

Most of the interest in this conference is focused on the conditions that enable the adult members of the public to increase or maintain their understanding of science, and it is with that outcome that I am primarily concerned here. Is there any evidence that school science courses help?

School science background and amount of use of informal sources

For an interaction between ideas from different sources to occur at all, the public with a background in school science must at least attend to the informal sources. Some evidence of the extent of the attention paid to science in the electronic media, the press and the scientific institutions can be obtained from surveys. Miller (1983) has some observations and I have reviewed some previous work, mostly from the United States (Lucas 1983). Here I wish to discuss some recent British findings.

In a survey conducted in the second half of June 1986, a representative quota sample (1033 people aged 15 or over) was interviewed in 82 constituency sampling points in England, Scotland and Wales. In addition to obtaining information about educational background, the support given to scientific research and views about scientists and scientific work, we asked a series of questions about the scientific knowledge of the respondents. These knowledge questions were mostly based on the level of understanding that could be expected of an average person who had successfully completed science courses at age 16 (that is, equivalent to a grade 4 in the Certificate of Secondary Education examination in England). The questions included a mixture of multiple choice, true–false, classification and free-response questions. From the answers to these questions a 'knowledge of science' scale was constructed.

Preliminary analyses of the survey give an indication of the relationship between scientific background and attention to various sources of informal scientific knowledge.

The press. The question asked about the press was

Which of the following subjects are you interested in reading about, either in newspapers or magazines?
Would you say you are very interested, fairly interested, not very interested or not at all interested in . . . ?

The responses for the question about science are shown in Table 1. On the surface, these results appear encouraging for scientists: 65% of the population report themselves as fairly or very interested in science news, which is almost as many as have the same degree of interest in medical matters, and a little more than the percentage interested in sport, although sport was by far the most frequently mentioned subject in the 'very interested' category (28%). This response rate for science is effectively the same as the 66% who expressed interest in media output in science in the 1977 survey of European

TABLE 1 Expressed interest in news topics (percentage of British adult population)

Topic	Very interested	Fairly interested	Not very interested	Not at all interested
Business and financial news	7	21	33	39
National politics	11	41	23	24
News about the Royal Family	15	41	23	21
Sport	28	34	19	18
Ecology and the environment	10	42	25	22
News about new scientific developments	17	48	20	14
Medicine	18	51	19	11
Religion	9	27	31	33

Source: MORI poll, June 1986. Sample size: 1033

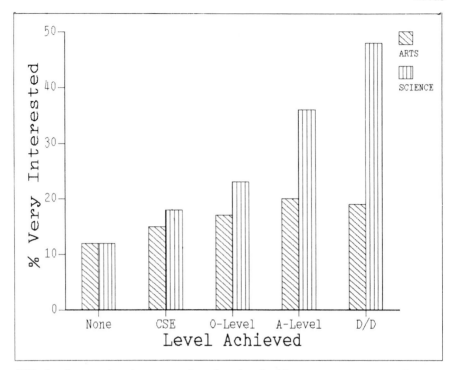

FIG. 1. Interest in science news by educational achievement: percentage of respondents who describe themselves as 'very interested' in news about new scientific
developments, by reported level of study. CSE: Certificate of Secondary Education, an
examination designed for pupils aged 16, and between about the 20th and 60th
percentile in general ability. O: General Certificate of Education, Ordinary level, an
examination designed for pupils aged 16 and in the top 20% of the ability range. A:
General Certificate of Education, Advanced level, a university entrance examination.
D/D: degree or diploma level. Arts and science categories are not mutually exclusive,
and many respondents reported studying both arts and science subjects at each level. In
the interview schedule, subjects were referred to as 'science' or 'non-science'.

public opinion (Commission of the European Communities 1977). However,
we are dependent on respondents' own definitions of what is a 'new scientific
development', and in the present survey 75% described astrology as scientific, and 30% saw cookery that way.

There are some interesting differences in the pattern of interest in science
news between those with arts and those with science backgrounds (Fig. 1).
There is a marked divergence in the expressed interest between these groups:
those with a science subject studied beyond O-level are much more likely to
express an interest in science news. In no other subject area did such a
marked difference appear. A majority of respondents at all educational levels
are at least interested in reading about new scientific developments, but there

is a selectivity associated with the level reached in formal science education. Similar patterns have been found in the United States (see Grunig 1980 for a review).

Television. Use of science programmes on television was investigated in a different way. We asked

From this list, which programmes would you say you make a particular point of watching?

The natural history programmes *Survival* and *Nature Watch* were viewed regularly by relatively more of those without a formal educational qualification. Other science programmes were viewed by a higher proportion of those with science than with arts qualifications beyond O-level, a pattern similar to that found for the press (Table 2).

TABLE 2 **Television programmes watched regularly, by educational level**

TV programme	Level(s) of study reported: number (and %) in category[a]								
	Science				Arts				None of these
	CSE	O	A	D/D	CSE	O	A	D/D	
Horizon	42 (26)	69 (26)	30 (33)	23 (33)	64 (28)	75 (23)	30 (22)	21 (24)	77 (19)
Nature Watch	31 (19)	60 (23)	14 (15)	16 (23)	54 (24)	61 (19)	28 (21)	17 (20)	146 (35)
The Natural World	33 (20)	63 (24)	20 (22)	23 (33)	43 (19)	77 (24)	32 (24)	20 (23)	115 (28)
Newsnight	26 (16)	46 (17)	16 (18)	24 (35)	36 (16)	57 (18)	28 (21)	24 (28)	85 (21)
Panorama	22 (13)	36 (14)	22 (24)	19 (28)	39 (17)	46 (14)	27 (20)	21 (24)	73 (18)
The Sky at Night	13 (8)	15 (6)	8 (9)	9 (13)	10 (4)	15 (5)	5 (4)	5 (6)	21 (5)
Survival	52 (32)	64 (24)	24 (26)	19 (28)	76 (33)	77 (24)	29 (22)	19 (22)	173 (42)
Tomorrow's World	70 (43)	90 (34)	42 (46)	33 (48)	98 (43)	113 (35)	46 (34)	30 (35)	110 (27)
TV Eye	21 (13)	34 (13)	13 (14)	9 (13)	37 (16)	34 (11)	16 (12)	12 (14)	82 (20)
World About Us	37 (23)	63 (24)	25 (27)	25 (36)	56 (25)	74 (23)	33 (25)	21 (24)	119 (29)
Number in category	164	266	91	69	227	321	134	86	412

Source: MORI poll, June 1986. Sample size: 1033.
[a] See Fig. 1 for an explanation of the four levels of education.

TABLE 3   Visits to science museums in last year, by educational level

Number of visits	Level(s) of study reported: number (and %) in category[a]								None of these
	Science				Arts				
	CSE	O	A	D/D	CSE	O	A	D/D	
None	108	152	56	38	151	205	85	49	342
	(66)	(57)	(62)	(55)	(67)	(64)	(63)	(57)	(83)
One	24	55	16	14	41	67	31	21	40
	(15)	(21)	(18)	(20)	(18)	(21)	(23)	(24)	(10)
Two	15	30	7	7	17	24	8	11	16
	(9)	(11)	(8)	(10)	(7)	(7)	(6)	(13)	(4)
Three or more	15	22	11	9	15	21	9	5	9
	(9)	(8)	(12)	(13)	(7)	(7)	(7)	(6)	(2)
Can't recall/no opinion	2	7	1	1	3	4	1	0	5
	(1)	(3)	(1)	(1)	(1)	(1)	(1)	(0)	(1)
Number	164	266	91	69	227	321	134	86	412

Source: MORI poll, June 1986. Sample size: 1033.
[a] See Fig. 1 for an explanation of the four levels of education.

Museums. Interviewees were asked

How many times, if any, have you visited a natural history or science museum in the last year?

The responses are shown in Table 3. There are no marked differences between those with arts and those with science qualifications.

In summary, there is some evidence that the level of formal science background is associated with attention being paid to potential sources of science information, and that those with higher arts qualifications do not attend to these sources to the same extent as their science counterparts.

Science knowledge and the use of informal sources

When we compare 'low', 'medium' and 'high'-scoring groups formed on the basis of the results of the knowledge questions, we find a strong association between level of expressed interest in science news and knowledge score (Table 4), with more of those with a high score being very interested in reading about new scientific developments. For television programmes, the relationship depends on the programme (Fig. 2). Similar patterns to those found for interest in reading science news occur for *Horizon*, *Sky at Night* and *Tomorrow's World*, with a higher proportion of those with high knowledge scores claiming to be regular viewers. For the other 'science' programmes,

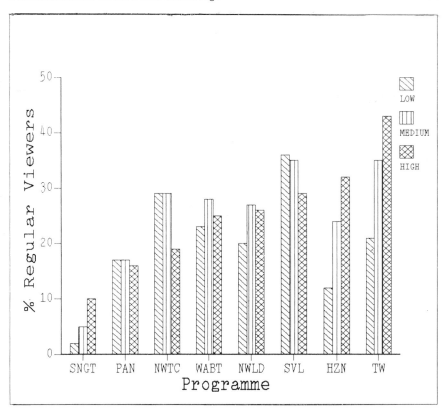

FIG. 2. Television programmes regularly viewed, by science knowledge score: percentage of respondents at each knowledge level who report that they 'make a particular point of watching' specified programmes. Knowledge level was assessed by a 24-item scale, and grouped as low (0–7), medium (8–15) and high (16–24). SNGT: *The Sky at Night*; PAN: *Panorama*; NWTC: *Nature Watch*; WABT: *World About Us*; NWLD: *Natural World*; SVL: *Survival*; HZN: *Horizon*; TW: *Tomorrow's World*.

there were no significant departures from the expected distributions. In particular, there was no tendency for the viewing of natural history programmes to be associated with knowledge group. The frequency of visiting a science or natural history museum was associated with science knowledge scores (Fig. 3), with the regular visitor group containing more people with a high knowledge score.

Although there is greater interest in science news by those with a greater background in science, and those who express this interest have a higher knowledge score, these findings do not mean that the science background of the high interest group has helped them to use informal sources of knowledge. The knowledge questions are relatively basic, and many of the respondents could have learnt the answers at school; the use of the media could be

TABLE 4   Expressed interest in reading about new scientific developments, by score on knowledge questions

Level of interest in science news articles	Scientific knowledge score: numbers (and %)			
	Low	Medium	High	Total
Very interested	18 (9)	113 (16)	41 (39)	172 (17)
Fairly interested	75 (36)	368 (51)	51 (49)	494 (48)
Not very interested	52 (25)	147 (20)	12 (11)	211 (20)
Not at all interested	62 (30)	85 (12)	1 (1)	148 (14)
No opinion	3 (1)	5 (1)	–	8 (1)
Total	210	718	105	

(Note: Low score is 0–7; medium 8–15; and high 16–24 on a 24-item scale constructed from responses to knowledge questions.)
Source: MORI poll, June 1986. Sample size: 1033.

an effect of the better knowledge. The knowledge scores of those who claim to be very interested in science news, compared with the scores of others with equivalent educational backgrounds, are also being analysed. Although such an analysis will give us more information about the nature of the audience for science news, we will still not be sure of cause and effect.

In a study of the *Origin of Species* gallery in the British Museum (Natural History) Griggs & Rubenstein (1982) found that visitors who had studied biology to at least Certificate of Secondary Education standard were better able to reconstruct the 'story' of the exhibition, tended to spend longer in the exhibition, and stopped at more of the orientation panels. As the authors point out, 'all this tells us is that people need to bring the knowledge with them to fully understand the exhibition's conceptual arrangement' (Griggs & Rubenstein 1982, p. 49). Their comment reminds us that we do not know whether the previous knowledge helped the visitors to learn about natural selection from the exhibit, or only helped them to recall the structure of the exhibition.

Griggs & Rubenstein's work illustrates a general problem in any investigation of the use of knowledge derived from one source while people are learning in another context. Without an experimental design comparing the scores of users and non-users of the exhibit we cannot be sure whether those with a better background learnt more. But if we do conduct such a study, by random assignment of visitors to the treatment groups, we are destroying the

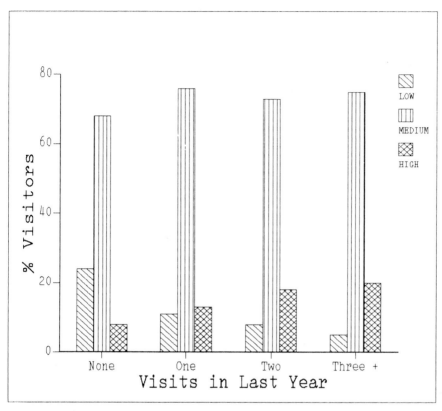

FIG. 3. Frequency of museum visits by science knowledge score: percentage of respondents in each visit category who scored low, medium or high on a 24-item test of science knowledge.

ecological validity of the informal learning setting, which is characterized by the learner making 'free choices' about the informal source.

There are, however, techniques that allow us to study learning-in-progress. P.A. McManus has been audio-recording groups of visitors conversing in front of selected exhibits in the British Museum (Natural History). Her analysis is not yet complete, but there are some instances where transcripts show the visitors struggling to use previously learnt knowledge in the context of an exhibit. For example, we give a transcript from two boys working at a classification game in one of the exhibits in the *Origin of Species* exhibition (Lucas et al 1986). The boys are clearly trying to impose what they remember of taxonomy on the specimens in the game, a strategy that may not be entirely successful at that exhibit.

By the nature of the medium, such studies of interactions in progress are much simpler in institutional settings: the observer can collect information while the interaction is taking place. It would be very difficult if not impracticable to do so systematically for television and the press. Much of the work is therefore likely to continue in museums, zoos and science centres, perhaps over-emphasizing the importance of these sources compared to that of the mass media.

Future research?

Should we continue with research into interactions between formal and informal sources of scientific knowledge? If we are seeking a well-developed account of how an individual's sources of knowledge interact it is difficult to go deeply enough into the formal contexts of learning, with the possible exception of museums, zoos and similar institutions. Without a well-developed account it may be difficult to use the results of the studies to guide the producers, writers and exhibition teams in their work of communicating science to the public. However, it is possible to provide more information on the actual audiences for each of the media by conducting studies similar to the survey described above. Knowledge of the audience may help.

There is potentially useful research to be conducted by extending the investigations of children's scientific ideas. As I mentioned earlier, this growing research tradition, which began with physics and is now extending into chemistry and biology, explores patterns of beliefs about scientific concepts. A number of common beliefs about dynamics, energy, heredity and light have been shown to occur among pupils in many countries. It is sometimes possible to detect the origins of specific pieces of information when interviewing pupils. Some pupils claim to have learnt about the structure of skeletal joints from television sports programmes: when an athelete is injured there is often a descriptive commentary accompanying a slow-motion replay of the incident. If more attention is paid to the sources of the pupils' ideas, we may begin to get some additional insights into the way in which ideas interact. But we need to extend beyond the descriptive documentation of 'alternative frameworks' that has characterized this research to date.

It is perhaps time to consider more carefully the psychology of learning in our studies of the use of informal sources. We need a psychology of learning in the context of everyday experience, in learning from experience of the commonplace, as well as from the contrived experiences offered by the school and the museum, and from the vicarious experiences offered by television and the press. Claxton (1984) draws on traditional psychological theories and applies them to everyday life in a way that may suggest fruitful approaches to researchers.

To understand interactions between sources we need studies of people learning, not just of the results of their learning. We need to look at individuals, not statistical aggregates. We need to be alert to heterogeneity, and not assume a uniformity. We need facts, not pronouncements.

Acknowledgements

The fieldwork for the survey reported here was conducted by MORI and supported by a grant from the Nuffield Foundation. I am grateful to Peter Hutton and Simon Braunholtz of MORI for their contribution to the survey design and analysis.

References

Bell BF 1985 Students' ideas about plant nutrition: what are they? Journal of Biological Education 19:213–218
Bodmer WF, Artus RE, Attenborough D et al 1985 The public understanding of science. Royal Society, London
Booth N 1979 The role of science in education. In: Reay J (ed) New trends in integrated science teaching. Unesco, Paris, vol. 5:13–21
Claxton G 1984 Live and learn: an introduction to the psychology of growth and change in everyday life. Harper & Row, London
Commission of the European Communities 1977 Science and European public opinion. CEC, Brussels
Driver R, Guesne E, Tiberghein A (eds) 1985 Children's ideas and the learning of science. Open University Press, Milton Keynes
Griggs SA, Rubenstein RW 1982 Orienting visitors to a thematic exhibition. British Museum (Natural History), London
Grunig JE 1980 Communication of scientific information to non-scientists. In: Dervin B, Voight MJ (eds) Progress in communication sciences. Ablex, Norwood, NJ, vol. 2:167–214
Lucas AM 1981 The informal and the eclectic: some issues in science education practice and research. Chelsea College University of London, London (Inaugural Lecture)
Lucas AM 1983 Scientific literacy and informal learning. Studies in Science Education 10:1–36
Lucas AM, McManus P, Thomas G 1986 Investigating learning from informal sources: listening to conversations and observing play in science museums. European Journal of Science Education 8:341–352
Miller JD 1983 Scientific literacy: a conceptual and empirical review. Daedalus 112(2):19–48

DISCUSSION

Laetsch: Recently Sam Taylor, who studied visitors to the Aquarium of the California Academy of Sciences, found much the same as you did—that people, especially those in groups, use exhibits in museums as a trigger to talk about something else in their lives. This must happen a great deal.

Falk: Many people go to museums for social purposes. The museum is fulfilling the mission many of its visitors ascribe to it (i.e. social), but not necessarily its more serious mission (i.e. 'education').

Duensing: I don't see anything wrong with that social aspect going side by side with the more formal principles of science.

Friedman: The National Science Foundation and other funding bodies may not think that paying for exhibits in museums is the best way to secure discussions of family history. If that kind of socializing is really one of our major functions in museums, perhaps we should be doing something quite different to promote such discussions.

Falk: It is probably presumptuous to assume that museums are the best way for people to learn about family history or science or anything else. An exhibit may elicit a response about family history from one group but a response about the behaviour of fish from the next group. How can we say whether we have failed or succeeded unless we have a much broader definition of success or failure?

Friedman: We used to provide science education simply by presenting the information and hoping it would get through. But children interpret sentences such as 'Animals are not able to make food' in different ways, depending on how they understand the word 'make'. To some it might mean the same as in 'My Daddy makes a lot of money'. Once an educator or a textbook author has heard that interpretation he or she is alerted to the need for a careful choice of words. A small amount of continual feedback is more important than having to define our purpose more clearly or having a Piagetian or other theory to start with.

Gregory: That is a very important point. Language comes from shared, not individual, experiences. In museums, particularly interactive museums, people can look at the exhibits and then the language of zoological classifications, or whatever it may be, makes sense. If they look at gyroscopes and then talk about force, the word 'force' takes on a rich meaning which can be shared by others who have seen the exhibit.

Science in schools needs to develop a language which doesn't depend simply on dictionary meanings but refers to the world of experience. This is the great difference between the arts and science. Science tries to make one interact with phenomena in such a way that the language and the method of testing a statement become objective and nearly everyone will agree once they have gone through this process. This is not the case with the arts.

The experience of science in schools is very important. The phenomena, the methodology and the sense of classifications provide the necessary framework for discussion. The living language based on experience and testing ideas allows one to be critical of other people without attacking them as people. Our human experience is enriched by discussions of this kind about the world around us.

Laetsch: The problem, then, is how people who have had this kind of experience communicate it to others who have not had the same experience.

Miles: The domino model of science communication postulates that you present the facts and theories, the knowledge domino falls over and knocks over the attitude domino, which then knocks over the behaviour domino. As we now know, this is a very poor model of communication. But it is used as an argument for science literacy. Much of the discussion of the Royal Society's report in the last 18 months seems to me to follow the domino model. I don't think, therefore, that Dr Laetsch was knocking down a straw man in his paper earlier today. If factors other than the presentation of facts and theories are important in communicating science to a lay audience and getting people to behave more knowledgeably vis-à-vis scientific issues, then scientists must acknowledge them not in the small print, so to speak, but in a way that properly brings out their importance.

The Royal Society report (Bodmer et al 1985) tends to give the impression that if one provides the facts, people will become knowledgeable about science and the nature of scientific activity. This will change their attitude to science and change their behaviour. What we have been hearing shows that this model is not sufficient. Walter Bodmer acknowledged earlier that there are other elements in the whole business of getting people to take an interest in science and behave as responsible citizens in a scientific society. Arthur Lucas and Beverley Bell have shown that there are—only recently acknowledged—educational, as well as social and political, factors to be taken into account.

Wolpert: Knowledge of science is just one of the factors that determine how we behave, and having a little knowledge is better than having none at all. But one should not be over-optimistic about what this knowledge will achieve. It is worrying that we treat science as uniform. Some people love biology and hate physics and mathematics, yet all these are science. There are also different styles in science—we all have different approaches to it.

If we had used the word 'politics' instead of 'science' in everything we have said so far at this meeting, would we find the idea very different? And what is the justification for art museums? In a way it is the same as for science—we think people will understand the world better and behave better. So is what we are saying special to science or to a particular intellectual attitude to life?

Lucas: The difference between what was in the Royal Society's report and what the public readers would say about the message of the report is the point at issue. It is not just the objective content of a communication that matters. It is its impact as a whole.

I want to emphasize that knowing the science is not enough to explain phenomena when people are confronted by new situations, and it is not just school-leavers who have to be able to change their minds. In our teacher-training programme we recently had a good example of this. Three science graduates were asked to plan an investigation to help children understand the origins of the film of condensation that appears on a beaker of water as soon as it is placed over a bunsen flame. They decided to heat a beaker of water on a hot plate. But what surprised us was the way in which they explained the absence of a water film in these conditions. They believed it did not appear because they had not heated the beaker at a sufficient rate. It turned out that they had believed the film was due to condensation of water vapour already present in the air in the laboratory, and they had saved their explanation by seeking to explain away the absence by a feature of their technique. People work very hard to preserve their own interpretations.

Caravita: The models of reality that people have are built up in different contexts than the scientific models, and they aim at different purposes. People do not change their concepts simply, when more evidence or more examples are advanced by experts. Common-sense knowledge and science are on different planes—they may interact, but they must be looked at in different ways.

Miller: The Royal Society report (Bodmer et al 1985) and the NSF report (1983) say that to be effective for large numbers of people, a solid science education is needed as a basis. Other experiences that people have may enhance their understanding, but these are no substitutes for a good science education. In the United States, at least, science in television, magazines, museums and so on serves the needs of people who have already had a solid education in science. Television and other informed sources do not replace basic science education, but enhance it around the edges.

Lucas: In the United Kingdom the level for a 'solid education' in science seems from my data to be at about A level.

Screven: Underlying a lot of this discussion are different views on the goals of science education—what science educators should communicate about science. What are the appropriate 'messages' to be communicated in classrooms, science museums, science publications? Are they systems of science? Science as a process? Factual knowledge? Problem solving? Should they include *affective* outcomes such as improving interest in science or attitudes about science? And what about the sequencing of goals? Perhaps there should first be a focus on enrichment experiences which prepare and encourage people later to take up science topics. I doubt if we can agree on such goals here. Yet, basic differences in what science education goals should be, and for whom, underlie a lot of our discussions so far. The disagreements I have heard cannot be resolved without considering the different goals of science education strategies. And these, in turn, depend on the particular deficits or misconceptions of a given target audience. And once we have decided what we want (or need) to communicate about science, how can we do this for a given audience, and can these goals best be achieved in formal settings or informal settings, or both?

But given acceptable goals and a teaching setting, these alone do not guarantee success—no matter how accurate, logical, and visually attractive the communication media may be. This is true in the informal settings of museums and zoos, on which I can speak most knowledgeably. Subject-matter specialists who plan exhibit content usually cannot, by themselves, prepare effective instructions, explanatory text, diagrams, or even content. In public settings, there are too many confounding variables that limit the effective use of learning theories or design models to predict how a given exhibit organization and visual framework will attract, hold, or communicate to voluntary audiences—except the 4 or 5% already motivated and knowledgeable enough to make good use of existing content (see Miles 1986).

To produce effective science exhibits, current evidence suggests that it is important to pre-test ideas for exhibit content and design with target audiences *early* in the planning process. This is because various details are very likely not to work without adjustments. If changes will be needed, it is important that these are made before it is too late and too costly to make them. Formative evaluation pre-tests exhibits during successive stages of development; one makes educated guesses on formats and content for exhibits and labels, then tests low-cost mock-ups of these with target visitors, changes these mock-ups as necessary, and retests until acceptable interest, usage and learning are obtained. This usually takes only a short time—a few days or even hours. The method is reported to be cost-effective by those who have applied it. Descriptions of exhibit applications can be found in Friedman et al (1979), Griggs (1981), Screven (1976, 1986), and Shettel (1973), among others. Griggs & Manning (1983) examined the validity of the method and report that it improves the post-installation effectiveness of exhibits.

Stewart: As someone who dropped out of formal science education at the age of 16 may I caution against treating the lay recipients of science communications as patients in need of treatment?

What is more important for many people is not that they receive a formal science education but that they should believe the communications about science that come their way. Why did the child Professor Lucas heard in the zoo believe the zoo's label that said 'Insects' rather than believing his teacher? I used to treat my teachers as people who could say no wrong!

Credibility is very important. The bits of science that come into the public domain in a big way are often the disasters, such as Chernobyl, Seveso or Challenger. People then want accurate information about that disaster. They may know nothing about the background science of atomic power, dioxins or space shuttles and it is too late to learn it at that stage. But they do need to believe the person who is explaining the scientific background to them. I have no solution to this problem but we all need to address the question of credible communications.

References

Bodmer WF, Artus RE, Attenborough D et al 1985 The public understanding of science. Royal Society, London

Friedman A, Eason L, Sneider GI 1979 Star games: a participatory astronomy exhibit. Planetarium 8(3):3–7

Griggs SA 1981 Formative evaluation of exhibits at the British Museum (Natural History). Curator 24(3):189–202

Griggs SA, Manning J 1983 The predictive validity of formative evaluations of exhibits. Museum Studies Journal 1(2):31–41

Miles RS 1986 Museum audiences. International Journal of Museum Management and Curatorship 5:73–80

NSF: National Science Board Commission on Precollege Education in Mathematics, Science, and Technology 1983 Educating Americans for the 21st century, vols. 1 and 2. US Government Printing Office, Washington, DC

Screven CG 1976 Exhibit evaluation: a goal-referenced approach. Curator 19(4):32–41

Screven CG 1986 Exhibitions and information centers: principles and approaches. Curator 29(2):109–137

Shettel H 1973 Exhibits: Art form or educational medium? Museum News 52:32–41

GENERAL DISCUSSION

Reaching the non-attentive group

Evered: The distinction which Jon Miller draws between the attentive group and the non-attentive group is very helpful when one is speaking about problems of communication. So far we have mostly been talking about the attentive group and ignoring the others. Yet the number of non-attentive individuals is far greater than that of attentive individuals, even if we don't have exact figures for countries other than the USA. We all seem to agree that there is a good prima facie case for improving public understanding. Obviously this would help the attentive people but what do we do about the second group?

The non-attentive individuals account for at least 75–80% of the population and the only inputs to this group during adult life are entirely accidental and random, derived perhaps from reading, as Arthur Lucas described, with people picking up snippets of unrelated information. What is our goal for this non-attentive group? Is it to reduce the proportion of non-attentive individuals and move them into the attentive population? Or are we merely going to try to improve their general level of education so that the non-attentive individuals will then have a better knowledge base and will be able to inform themselves, if motivated, on the basis of personal interest?

If we convert some of the non-attentive into attentive individuals what are we going to do with the residue, who will probably still be the larger part of the population? We all probably feel that they ought to have a better knowledge base for understanding issues relating to their health or other matters in their lives. We perhaps ought to draw a distinction here between the appropriate background and formal training for those who want to maintain their level of general education in the sciences throughout their lives and those whose interest will be limited to finding out about and understanding specific items of information when they perceive the need to do so.

Screven: The problem with the attentive population is not their lack of interest in science but their misconceptions or preconceptions about science. The attentive population probably includes many adult museum visitors so science museums may be good places to correct their misconceptions/preconceptions and improve sensitivity to fundamental issues in science. Preconceptions can have an enormous impact on how people perceive the world, how they process information and what they attend to. In science centres, one can interrupt this process by encouraging overt interaction with exhibit content— using, for example, leading questions in headings and in labels that direct attention to conflicting evidence or provide devices that encourage viewers to act on common misconceptions, with results that lead to the use of alternative interpretations (Screven 1986). If leading questions, computers, and other activities with visual and interpretive exhibit content were coordinated, attentive individuals could be brought into direct contact with experiences that focus on science as a process, as a way of thinking, and with basic issues of verification.

Science exhibits as now constituted often have the opposite effect: they reinforce notions of science as technology, as anything factual, and so on. They contain more language-laden content than they probably should, given the leisure-oriented nature of most visitors. Museums probably are *not* good places for in-depth and vocabulary-laden science information. (They may be more appropriate for 'right-brain' processes than 'left-brain' processes.) Visitors are not in museums long enough to learn or retain much specific information during any one visit. And the time spent at any one exhibit (or topic area) is usually too short for visitors to acquire in-depth knowledge. In informal settings, it seems more reasonable to aim to improve attitudes, stimulate interest, provide models for approaching everyday questions scientifically, provide models for ways of asking questions, encourage divergent thinking, and replace poorly conceived beliefs about science with more appropriate ones. It seems reasonable to assume that even a small impact on such matters would affect how viewers are likely to approach related issues and topics outside the museum and, one hopes, increase the chances that science will ultimately be better understood.

Quinn: Infrequent visitors do not go to museums specifically to learn or become more cultured. They go for social purposes or for family activities, as we heard earlier. Museums need to develop communication techniques that address the social reasons for visiting museums. We need to capture people's interests and dispense with some of the formal approaches that we have been trying to impose on informal environments. Presentations that allow for social interaction will help to promote further understanding. Visitors don't need to know they are going to learn new facts or acquire new knowledge; instead they should get a whole new feeling—new enthusiasm and excitement—about science.

Evered: Arthur Lucas told us that more than 60% of people do not visit museums. How do you capture the interest of those people?

Falk: One perception in the public's mind, reinforced by the formal education systems, is that the words 'learning', 'education' and 'school' are somehow synonymous. Another perception is that learning is roughly equivalent to what you do in a classroom. Both of those perceptions are counter-productive to the goal of improving the quality of science literacy in the general population. Those two perceptions conspire against cooperation between the formal system and the informal system, against an expanded conception of learning which allows people to feel good about doing things other than memorizing facts when they walk into a museum. How do we break down those stereotypes and myths which get in the way?

Quinn: And how do we who work in museums learn to present material to the visitors?

Thomas: Have we clarified *why* it is important to enhance the 'understanding of science' by the attentive or non-attentive groups? It is not enough to say there is a prima facie case. Is it for their personal enrichment, to make practical decisions in their lives, or is it to benefit the economy?

Laetsch: There is probably a general agreement that scientific literacy is necessary but not sufficient. We all agree that it is a good thing and that people ought to know something about science. What it leads to is something else.

Macdonald-Ross: No case other than old-fashioned evangelism has been made for there being any positive consequences of getting the 60% who don't visit museums into museums. You could argue that reaching them is impossible anyway.

Laetsch: The non-attentive people are not always non-attentive. They are a shifting population, not archetypal morons.

Miller: That is central to the argument I was trying to make. The composition of the attentive public may change from time to time. If the political system is to perform adequately, particularly at times of crisis or controversy, the attentive population needs to be well informed. That is a persuasive case for the 20% who are attentive.

The issue of what to do with the non-attentive population is extremely important and we need to decide some priorities. If universal scientific literacy is a desirable objective we probably know how to achieve it, but are we prepared to pay the cost? I suspect that society will not be prepared to pay that cost for many years. Science education will always operate under limited resources. Given that constraint, what do we do to enhance scientific literacy for both the attentive group, to which I personally would give high priority, and the non-attentive group? My answer would be to build a good secondary science programme so that people can follow science issues if they choose to do so. By not providing that opportunity at high school we deny some people the choice. Out of that choice will come audiences for museums, magazines, television shows, and so forth. We must not ignore the non-attentive individuals. The least we can do is to provide a common education which gives people options.

Thomas: But you should work out why you want to do it before you work out how to get to that point.

Wolpert: Science is difficult. It simply is not true that it becomes easy and interesting if you know how to do it. The scientific mode of thought is an alien mode for most of us; it is uncomfortable, unnatural and very difficult. We must face that before we understand the nature of this problem.

Laetsch: Jon Miller defines attentive people as being informed enough to participate in issues related to public policy. Large numbers of those you call non-attentive are in fact attentive to various aspects of science and technology. In the USA a significant percentage of the adult population has enrolled in continuing education courses, many of them science-related. It is not television and science magazines that catch people's attention so much as the hobby magazines, the lay technical magazines in all kinds of subjects—electronics, mineralogy, gardening, bird-watching, nature study, environmental concerns, and many others. People are interested in these for their own personal goals but at various times that interest may lead to their concern with public policy issues. Bird-watchers and others will definitely get involved in something which relates to preserving a habitat for birds, though they may not care anything about other topics.

Hearn: On the one hand we want to get across that science is part of everyday life. It is there all around us and we want people to have a good feeling towards it. On the other hand we don't want people to be too comfortable with it, because if scientific progress is all so simple, why do we need to pay so much for it? This is a dilemma. The attentive group and the decision-makers need to appreciate far more of the research process and this needs to be tailored more for them than for the non-attentive group.

Deehan: Most of us here have a vested interest in this question because we are mostly employed in some way to communicate science or provide information about it for people who wish to have it. We feel that everybody has to be interested in science and educated in it and have it communicated to them. But there is no law that says everybody has to be interested in science. If people want historical information or information about art or politics it is available to them. Science is the same. It is provided for people and if they want it they can take it.

Rhodes: But are we earning our money if 95% of the population is not being reached? Our discussions might be more fruitful if we looked on ourselves as failures. Are we being too complacent?

Nelkin: Nobody has yet mentioned the general level of illiteracy and the fact that a lot of people do not read. Where does science education fit in this kind of context?

We are talking about science education and communicating science as if science is inevitably neutral, totally positive and completely benign. But how do we communicate information about the potential dangers of science, such as its contribution to militarization, or the eugenic possibilities that are inherent in new biological research? Is there any effort to train people in genuine citizenship and give them a sense of the social and political consequences of science. All the discussions so far have been plain promotional.

Friedman: I am humbled and somewhat dismayed. At most meetings I go to we discuss the efficacy of a particular technique for accomplishing something. We started out with a bang here by questioning the very existence of the enterprise, what its aims are, whether it is worthwhile, and whether it has any efficacy at all. Mac Laetsch started it but Jon Miller's stunning figure of 5% scientifically literate people is what we have been responding to. Even if we question the definition, that figure has gone down in the last 20 years, even though in the USA in that period we created about 100 new science and technology centres, spent lots of money on curriculum development, and poured money into television programmes on science.

Today we very quickly got down to scratch and started wondering where the enterprise is going and why. Most scientific conferences deal with fairly well defined questions, like 'How did that protein get there?', and the debates are on whether the answer is known yet and, if not, what the best technique is for finding it. This conference, in contrast, is much more like a humanities symposium in which we question whether everyone needs to know Shakespeare and whether we can explain Shakespeare to people who can't read or write. We are starting off with a position of great openness and defensiveness which surprises me to some extent.

There is an undercurrent of some optimism in response to the question about reaching the inattentive public. All of us seem to feel that a little bit of science is good for everybody. First, most of us are paid out of the general revenues of our society and feel we have some obligation to justify that by communicating what we do to the broad masses who pay the bills. Secondly, we just love science. We think it is beautiful, we think it is spectacular that we can peer into a box and see something that has not been seen before, and gain an understanding of something that had not been understood before. This latter is an evangelical mission which has no further defence than that we love science and have an emotional need to share our love with the rest of the world. Those two arguments may be more pertinent than our particular arguments over efficacy, since the data for efficacy seem dismal.

Whitley: Earlier I said we needed a clearer focus on who was communicating what to whom, and why, and with what consequences. That is still a useful set of questions to ask in a meeting dedicated to the communication of science but I would now like to discuss the whole issue of applications. One of the arguments for scientific literacy is that somehow one is producing or communicating a skill akin to literacy. But that is extremely dubious. Scientific research and the communication of scientific results to other people in the development of new research is an esoteric and unusual activity which relies on highly coded methods of proceeding and communicating which are alien to everyday practical discourse. Why else does it take seven years to train someone to be a post-doctoral research worker? If you are trying to train someone in practical reasoning or argumentation, it can be done much more quickly and cheaply than by educating them in formal science. It is a skill you are trying to train them in and if as a democracy we need that skill of reasoning things through from the information that is available, we don't need to spend all this money on communicating science in the conventional sense of the term. Creating scientific literacy is not essential to creating a skilled capacity in the sense in which teaching the ability to read and write is teaching a series of skills.

And what is it you are communicating? Is it an eternally true sort of knowledge? That is often the image people have of science, and scientists like it because it justifies all the money we have paid so that they can go on adding new bits of knowledge which are also eternally true, and so on. If it is a process that is being communicated, which process are we talking about? If it is a process for practical reasoning, you can teach a lot with the appropriate set of conventions which we happen to have reified into logical inference procedures. Or do you want to teach people how to manipulate cloud chambers in particle physics, in which case a much longer process is needed—but there are quite enough particle physicists already.

These questions need to be addressed. If this conference is meant to produce a book to be read by presumably over 1000 people, we have to think more seriously about these questions than we have done so far.

Lucas: I want to reinforce the questions of what, why, where and when. One issue we discussed was the role of the formal curriculum in schools. It is very difficult to decide what base we are building on. If we focus on the process we may replace one set of dogma with another. The way we teach the method of science in schools tends to become a set of rules to be followed unthinkingly. What bits of science should we build into our school science courses? I suspect it is the four or five large generalizations of science which make sense in most fields—such as the particulate nature of matter, the principle of conservation of energy, ecological interdependence—but then a lot of the fine-grain detail would be lost. Some of the detail would be needed to help to develop the understanding of the large generalizations, but that would be as a means, not an end.

We are skating around that issue, but we are also skating around it in terms of informal sources of learning. We talked about 5% of the population being scientifically literate but the definition of scientific literacy we were using focuses on civic responsibility. The chairman reminded us about the bird-watchers and other hobbyists. They use a focused set of scientific information and a particular set of tradition or good practice. Scientific literacy for civic responsibility and the use of science in hobbies don't necessarily coincide. We believe we can help the practical skills and development with particular bits of science. But that makes it even more difficult to know what to put into school science because we are trying to support a wide range of specialist interests, not just improve the discussion of science policy. We should think more carefully of the multiplicity and heterogeneity of our population instead of regarding the population as being uniform.

Gregory: Should politicians themselves exhibit some knowledge of science? To get rid of bugs in a large computer program you need another computer which looks at all the possibilities—essentially a simulator. You cannot make the simulator work more than about three times as fast as the computer, therefore it is absolutely impossible, in the life of the universe, to consider all the possibilities of a large computer system. That type of logical argument never comes up in political discussion but it is vital for example to the Star Wars idea. If something has bugs in it you can't rely on it, therefore it is not a proper system. And it is incredibly dangerous because it could make itself go off.

I think politicians don't talk about this because they don't understand it. Educated people have a real dislike of politics and distrust politicians because they so obviously don't exhibit scientific understanding. The distrust will grow, until either politicians buck themselves up or we all become anaesthetized. Surely part of what we are trying to do is to set up an educated community in the press, television and radio which politicians listen to. This debate can then be carried on at a higher level of decision-taking by politicians and by the general public, as we are a democracy. Unless we can have a debate at this sort of level, what is democracy in a world which has technology? I submit that what we are doing is vitally important if we are to have a technological world.

Whitley: You can't get away with it as easily as that. You have to specify what sort of knowledge enables people to become more democratically acceptable.

What sort of knowledge are you communicating to politicians? How do you justify that bit of knowledge rather than another bit of knowledge, or that bit of reasoning rather than another bit of reasoning?

Gregory: I was giving the reasons why we are here. Some people were worried whether we are pushing our own interests because we are being paid to do so. But there are deep reasons for this kind of activity. I am not giving any answers, just setting up the premises.

Laetsch: We like to think that because we ourselves operate on a logical premise of some kind, other people do as well. Hence the Star Wars analogy. A lot of people think that Star Wars has little to do with science but is purely a political ploy. If we can get involved with it, the Russians will follow and their economy won't be able to stand it. That is taken seriously by a lot of people. A rational or scientific argument against it has no validity because that may not be what it is all about. Scientists feel that everything has to be debated in our terms, and it is only through our terms that decisions will be made or some enlightenment accrue. When we face that particular question we don't have many models to go by. Science educators have not given it a lot of thought.

Nelkin: During the Kennedy administration there were a great number of scientists serving as advisers in government and it was often a political disaster. I am not sure that scientists running the country would be a good thing. Politics does not exactly operate according to scientific rationality.

Gregory: I am saying that this is a necessary, but not a sufficient condition.

Laetsch: Our lack of success so far in formal education, and perhaps even in informal education, may be because our definitions of how we learn science and where we want it to go define the way in which we communicate it. Perhaps it is not a viable system.

Reference

Screven CG 1986 Exhibitions and information centers: principles and approaches. Curator 29(2):109–137

Science broadcasting–its role and impact

GEOFF DEEHAN

BBC, Broadcasting House, London, W1A 1AA, UK

Abstract. Science is usually presented as a continuous, coherent, concrete activity. It relies for its insights on the systematic analysis of a mass of theoretical and experimental observations about nature. It is also hard. By tradition, science deals with the outside world through special vocabularies and armouries of arcane facts. Science does not like superficiality, trivialization, or talk of 'breakthroughs'. In its nature, broadcasting is episodic, fragmentary, ephemeral, and unable to aspire to encyclopaedic coverage of any subject. Its currency is the everyday language of the general public. This mismatch between the demands of science and the ability of broadcasting to meet them creates many profound problems in communicating the significance and excitement of scientific discovery to a lay audience. I explore these problems and consider ways in which scientists and broadcasters might overcome them.

1987 Communicating science to the public. Wiley, Chichester (Ciba Foundation Conference) p 88–99

I have quoted this story on several occasions, and I even stole it in print once. I make no apologies for using it again, but it gives me great pleasure to let the original story-teller tell it this time:

As I was coming here I heard the usual remark . . . that I'm coming to hear your lecture . . . of course I know I'm not going to understand any of it and so forth. I realised that when you talk of these scientific lectures the difficulty is to imagine these complicated things that you're not used to . . . it takes a lot of imagination . . .

Reminds me some years ago, I left the ivory tower for just a little bit—just a weekend or so, and found myself at a party . . . the apartment was on the Santa Monica Pier in the Building of the Carousel. So you get some idea of the type of circumstances. And there were a lot of these sort of art-like semi-intellectuals there and so on. And when they found out I was a scientist, I got into a lot of arguments, and I remember one argument about imagination.

We got involved and the man said 'We have the hard job. We have to imagine all kinds of things, many that don't even exist. But all you have to do is imagine what's there'. And he's right. That's all we have to do. Seems to be awfully difficult for us to do it. Seems like my life's work is trying to imagine what's there.

And it was the same guy of course who only a little bit later was explaining to me why he knew that Einstein's Theory of Relativity was wrong. Why? Because it's inconceivable . . .'

Dick Feynman is probably one of the best communicators of science there is. Yet, as his story implies, getting scientific information across to a lay audience is not usually an easy task.

The title of this paper is 'Science broadcasting—its role and impact'. I will be discussing impact later. Earlier in this conference we discussed some of the reasons why we want to communicate science. My own reason is that I work for the BBC. The BBC exists to disseminate culture. Science is clearly a very important cultural activity, and my job is to bring it to as wide an audience as possible. Now let me give you my own fairly simplistic definition of the role of science broadcasting.

It is to enable people with scientific insights into the way the world works to tell them to the rest of us.

Broadcasting has some unique strengths to bring to fulfilling this brief. But in unprepared hands radio and television are perhaps the most unforgiving of media through which to explain complicated and technical matters. This piece of transcript illustrates the fate of someone who is unaware of the demands of broadcasting to a lay audience:

If you look for example at many enzymes . . . to take a simple example if you take an enzyme which is a digestive enzyme, chymotrypsin, the enzyme has in its catalytic site—in the part which actually causes the cleavage of the peptide bond in the substrate—it has a combination of a histidine residue which is at position 57 in the chain, and a serine residue which is at position 195 so that if you were to string out the protein chain these would be approximately 140 amino acids separated in terms of a linear sequence.

The nationality of that gentleman is immaterial—the interview is simply the latest in a line I could play you, stretching back to the beginnings of science broadcasting. Part of his trouble was, of course, imposed by the nature of the activity he is involved in.

Science is usually presented as a continuous, coherent, concrete activity. Scientists spend their time assembling a mass of theoretical and empirical observations about nature. They then apply a systematic analysis to these observations. From the analysis, they extract insights into the way the world works. As Dick Feynman suggested, science is pretty hard. The truth is not won easily. Nor is it always as simple as we would like it to be.

Like every other profession, science has developed its own ways of communication. It deals with the outside world through special vocabularies and armouries of arcane facts. The value of any piece of scientific knowledge is ultimately bound up with the quantity and quality of the evidence supporting it. The footnotes can often be as important as the main text, and they must be included.

The reasons for this state of affairs are obvious. Unique vocabularies prevent ambiguity. The facts are obscure because the world is a complicated and subtle place. The detail is necessary to demonstrate that the view presented fits the constraints imposed by nature.

Sadly, the way scientists talk to each other is not at all appropriate for broadcasting. Please note that I am talking here about every form of science broadcasting, not just the sort of programmes made by my unit or Mick Rhodes' department.

Broadcasting is episodic and linear. Things are usually presented in discrete chunks, not as part of a continuous body of knowledge. They go past you in a stream of words and pictures. And although we could, given advances in recording technology, stop and play back parts of a programme, in practice we tend not to do so. The general public still regards broadcasting as an immediate and transitory activity. This is in complete contrast to a book or magazine, in which re-reading is an easy and profitable part of the acquisition of information.

Broadcasting is fragmentary and ephemeral. It deals by and large with image and anecdote. It is extremely difficult to present a line of coherent and comprehensible argument—especially when pictures are the primary means of presentation.

It is comparatively easy to make those pictures present dramatic and striking images—but images whose major impact is on the emotions. I have, for instance, often asked television colleagues to show me a picture of a moral dilemma. I am still waiting for one.

Broadcasting can never be encyclopaedic. Air time is restricted. It takes far longer to present a given amount of information in a broadcast than through almost any other form of communication—except perhaps conversation. This talk consists of just over 2500 words. It will take me about 20 minutes to deliver them to you. It would take you about 10 minutes, or less, to read them.

Television and radio producers are under constant pressure to fit their programmes into an allotted space. It is not surprising that the first things to be omitted are the details. To the producer the broad sweep is what matters. To the scientists, the context and supporting evidence are crucial.

It is also very easy when making a science programme to find yourself impelled by what can be called the 'Show Business Imperative'. This afflicts principally producers who are not themselves convinced of the intrinsic interest of the material they are dealing with. It is manifested in the employment of 'personalities' to present programmes, in the construction of 16-foot high models of the human nose, and in concentration on the trivial aspects of science. And trivialization, of course, is one of the things that makes scientists cross.

But worst of all, the language of broadcasting is the everyday language of the general public. We aim, as we must, at a lay audience, one sometimes measured in millions. The proportion of listeners and viewers in this audience who understand the specialized vocabulary of science at even a trivial level is vanishingly small.

You will have gathered by now that I believe that science and broadcasting are uneasy bedfellows. This is not to say that excellent programmes have not been made, nor that they cannot continue to be made. But my view is that the mismatch between the demands of science and the ability of broadcasting to meet them at present compromises the quality of much science broadcasting, in particular news and current affairs coverage.

What can we do about it?

The first thing we can do is stimulate scientists and broadcasters to trust each other more. An apocryphal story is told of the first time a radio producer went to see an esteemed scientist to persuade him to broadcast. The scientist agreed readily. But as the producer was leaving, the scientist caught him by the elbow. 'One last thing', said the scientist, 'what's the fee?' In trepidation, the producer mentioned some trifling sum. 'Fine', said the scientist, 'To whom shall I make the cheque payable?'

Those days are gone forever, and it sometimes seems to me from conversations with researchers that the scientific establishment regards dabbling with the media as an unclean activity. This is often revealed when we try to get younger workers onto the air. Too often, the shadow of their professor looms across them during the interview—listening for mistakes, indiscretions, or just their attempt to talk in language anyone can understand. Too often, the result is a stiff, fragile, even incomprehensible, performance.

There are slight signs that this is changing. Scientists are becoming more aware that they owe the public something for the opportunity to pursue their research. What they owe them, in my view, is the chance to share in the cultural excitement that scientific discovery provides. And one thing that broadcasting is good at is communicating emotion and mood—the elation of original discovery.

If science is to prosper, it must take advantage of every opening offered to it to persuade the public that it is something worth doing.

Paradoxically, I have found through extensive travel and interviewing in the USA that American scientists seem to have exactly the set of skills needed for radio broadcasting. They know the audience. They are not afraid to generalize. They tell jokes. Above all, they seem to enjoy it. It is a paradox, because America has no radio programmes comparable to those in the United Kingdom . . .

I acknowledge, though, that although microphones and cameras (at least as pointed at scientists, and sometimes *by* scientists) are usually benign beasts with honourable intentions, they can also bite. Again from conversations I have had, I know that scientists fear both the interview and the editing that follows. They fear that the essential qualification will be missed out, that the caveat so necessary to avoid raising false hopes will unaccountably end up on the cutting-room floor. They fear trivialization, imagining that everything they say, no matter how profound, will end up either as gee-whizzery, or as 'look what those crazy scientists are up to now'.

I have seen radio and television journalists in action with scientists in many countries of the world. The most charitable explanation I can think of for the behaviour of some of the journalists is that they spend so much time dealing with crooks and charlatans that they forget how to behave with honest folk. Anyway, the remedy for that sort of abuse is clearly in the hands of the broadcasters. As a group I do not think we take enough steps to ensure that the best of specialist knowledge from programme-makers intimately involved in science informs the activities of general-purpose journalists who work mostly in news and current affairs. We will try to do better.

The other thing that we can do is to spell out more clearly our aims and objectives, and the constraints under which we operate. There is much talk at present about broadcasting being the language of priorities. It must seem to anyone willing to do the sums that science has a low priority. For instance, I have three producers besides myself to cover the whole of science for national radio. I suspect that other broadcast editors would not complain of overstaffing.

In discussion among professionals, it is almost invariably agreed that communicating science by broadcasting is one of the most difficult tasks there is. Why, then, are we not overburdened with staff and resources? I regret to say that I have no answer. Perhaps you would care to write to your MP, Congressman, or whichever deity seems appropriate . . .

But given these limited resources, clearly I want to do the best I can with them. And here the scientists can help.

My aims can be stated quite simply. I want to bring to a lay audience discoveries and developments from the frontiers of science in a way which will engage their attention and excite their imaginations.

When scientists are good at explaining things, they can be very good. And it is usually the best scientists who are best at explaining. This is because they are confident enough of their own standing to step back and present their work in bold sweeps. They recognize that they are speaking to a general audience—and a general audience does not require footnotes and references to the past 16 issues of *Nature* to appreciate the worth of a piece of work.

What matters to the general audience is to be given the idea, the conceptual information that will illuminate what the scientist has been doing, and why it is important.

I am not arguing that supporting evidence is peripheral, but that it would help us all if scientists would decide which bits of evidence are really essential to understanding the story, and which bits of detail obscure the central message. The public understanding of science has been the subject of a great deal of thought. In fact one of the most distinguished thinkers on the subject is taking part in this conference. I am broadly in sympathy with the conclusions presented in Walter Bodmer's report to the Royal Society (Bodmer et al 1985). I think the task of all of us would be eased by better education in the sciences and by more sympathetic treatment in the popular press, for instance. There is one other conclusion I would like to add, though.

It is this: scientists do not spend enough of their time telling stories. And when they do, they all tell different ones.

Over the years, I have lost count of the number of different metaphors I have been given to explain any particular concept in science. Some have been good, some bad, but all have been transitory. I do not, therefore, find it surprising that the general public doesn't have in its head some bell which rings when relativity is mentioned, in the way that they have one which rings when television is mentioned. Notice that I am not saying that the average member of the public understands how a television works. All I am saying is that the word does not frighten him or her.

Why can't we perform a similar operation for all the bogey-persons of science—quantum mechanics, red shift, particle physics, neurochemistry, and the rest?

Consider this. We change our curricula so that a part of every undergraduate course in our universities consists of teaching the students the standard metaphors for the important concepts in their discipline—and in others, for that matter. When the students graduate and go out into the world as scientists, they use these metaphors to explain what they're up to to their mothers, children, and Joe six-pack in the bar in Wisconsin.

And since the metaphors are chosen for their beauty, elegance and wit, they engage the attention of the audience. They enter the public consciousness. In not too long a time, we find that talk of scientific topics no longer causes instant eye-glaze. The scientists are happy. The broadcasters are happy, because we no longer have to spend so much of our time on background. And the public are happy, because they can get past the conceptual barrier and appreciate the excitement—and value—of science.

My proposal, therefore, is that we assemble the best story-tellers in every branch of science, and lock them in darkened rooms until they come up with the stories we need. I even have a candidate for chairman and chief inspiration—Dick Feynman, talking here about the neutrino:

This particular subject I've chosen is going to be very easy for you. You're all going to understand it. I want you to imagine one thing. All you have to do is imagine something that does practically nothing. It does almost exactly nothing except exist . . . You can use your son-in-law for a prototype . . .

References

Bodmer WF, Artus RE, Attenborough D et al 1985 The public understanding of science. Royal Society, London
Feynman R 1978 Alumni Day Lecture. California Institute of Technology, Pasadena (13 May)

DISCUSSION

Laetsch: The reason you find American academics better broadcasters than British academics may be because they often teach courses to between 500 and 1000 people. They speak into a microphone and often face a television camera. They have to tell stories to keep the non-science majors awake at 8.0 in the morning. They are proto-broadcasters.

Broadcasting is like formal lecturing in that both are public performances. This may be why broadcasting is not a good way of getting science across.

Deehan: I agree. There is little interaction between lecturers and their audiences. Conversation is the best way to get detailed information across.

Laetsch: In the USA most formal science education is done through lectures. High-school teachers simulate the way they learnt science at university.

Tisdell: Direct contact and facial movements seem to help in communication.

Community ratio stations, sometimes operated by universities, are another route for science communication in Australia. Does that happen in England?

Deehan: Community radio tends to be very light-weight here but we have the Open University on national radio and television.

Evered: Since we began running the Media Resource Sevice at the Ciba Foundation we have been able to introduce a wide range of scientists to broadcasters and journalists. Many of the scientists who have since been interviewed on radio or television had no previous experience of these media, yet it seems they enjoyed the event and the feedback indicates that they were better interviewees than you suggested. The available resources are probably better than you think.

Deehan: I wasn't suggesting that the resource wasn't a wide one. In our programmes we try to bring new voices to the microphone, though in other areas of broadcasting the notion of finding spokesmen for science is still quite strong. The Media Resource Service is a good thing and I hope people in the radio and television services will look at how it might be used and at the lessons it provides.

Dixon: Radio and television in the UK do a better job of putting science across than national newspapers do. A key difference between a good interviewee and a bad one is their syntax. Scientists tend to use very boring syntax which is completely unnecessary. If you talk to someone over lunch, he or she will say 'We are beginning to think that A causes B'. The same person in front of a microphone will say 'It has been shown on preliminary data by X in Baltimore that . . .' Some people can adapt to their audience, some of them can't.

Scientists have to use certain technical terms but I don't think those terms are a problem. We are all imbibing new technical terms all the time—strategic defence initiative, zero option, Big Bang, and many others—and we should not need to make an exception for science.

Good interviewees, especially Americans, tend to talk in a straightforward way. But many British scientists slip into this awful syntax which they use in scientific papers and on platforms at meetings.

Deehan: It is impossible to change the way people have been talking for 20 years. I can only tell our contributors who the audience are and hope they will talk appropriately.

Rhodes: Television producers, who use written scripts, tend to pick up that nasty infection from scientists. Editors then have to spend time converting the scripts back into English.

Wolpert: Geoff Deehan said that radio was good at getting emotion over to an audience. But I don't think you get much emotion over in relation to science. Both TV and radio homogenize scientists so that they are indistinguishable from one another. They are seen as clean white-coated talking heads who never tell jokes and seldom laugh. A very stereotyped image is being perpetuated in which the variety of styles and personalities that exist in science are missing. This image bears very little relationship to the way science is done.

Nelkin: Actually there seems to be little humour in the scientific community as compared to the medical community. Scientists don't seem to have a well-developed tradition of humour or science-based jokes.

Wolpert: But scientists spend a lot of time laughing at the disasters that happen in the laboratory. It is more subtle than the formal joke.

Deehan: It was difficult until about five years ago to get scientists to talk about their mistakes. Now they are prepared to say why something didn't work and that they are trying again.

Stewart: That bears on the point I made about credibility. If people are prepared to expose their mistakes in public, their credibility leaps up.

Hearn: Perhaps one of the major problems is language. Scientists are trained to state a question, develop an argument and come to a conclusion. By the time they get round to developing the argument in an interview, time has moved on and the producer has lost interest. One needs to get the punch-line in first.

Deehan: The old joke about a joke is that you tell people what they are going to laugh about, then you tell them the joke so that they laugh, and then you tell them what they laughed at. You have to do the same in broadcasting, whatever people are listening to. But scientists don't do that.

Hearn: Details that are important to us as scientists are unimportant to producers. This morning I had an irate phone call from some co-workers abroad who had not been mentioned in a BBC radio programme in which I was involved. The subeditor had taken the reference out although I had stressed that those people must be mentioned.

Deehan: It is very difficult to deal with that kind of problem. A whole string of names is of no interest to the public. We can say 'John Hearn and colleagues' but some people don't think that is good enough.

Caravita: Television programmes on wild life have contributed a lot to changing people's attitudes. Apart from the capturing power of the images, zoologists and naturalists seem to have a different style of presenting themselves and their knowledge. In contrast, medical programmes present the results of research and the work of the scientists in a much more authoritarian way. Does this have anything to do with the status of these disciplines?

Deehan: The sorts of disciplines producers prefer to deal with start with natural history because everyone is interested in animals. Then comes medicine, because so many people are hypochondriacs and the programme contents are close to the everyday experience of the general public. Science comes third, because much of it involves an alien way of thinking and we have to convince people that they should spend time having their ideas about the world changed.

Laetsch: Should we make a distinction like that between science and natural history? The intellectual force in biology is derived from an interest in fuzzy animals and in plants.

Deehan: I was thinking of programmes about animals running across plains and killing other animals.

Gregory: Patrick Moore has probably done more for astronomy through his work in the media than any professional scientist. Do you need more people like that for the other sciences, with completely natural personalities? Do you have a policy of looking for such people?

Deehan: We certainly need good communicators who can raise the public awareness of science without making scientists cross. We look for people with real authority, for presenters who are practitioners right at the frontiers of science. I am not an expert in any of the fields of science about which I make programmes, so I need somebody who knows what important questions need to be asked. To plumb the depths of science one needs someone who knows that there are other questions and implications that must be teased out.

Rhodes: But David Attenborough, for example, would claim to be no more than a naturalist, never a zoologist. I would give my eye teeth for a scientific disc jockey or two with wide appeal but they don't seem to exist in the UK and perhaps not in the USA either. Jonathan Miller, for example, appeals to a quite restricted audience.

Evered: What is the size of the audience for science broadcasting and what is its composition? What lessons can you learn about the topics and formats of programmes that might help us to approach the question of how to interest a progressively larger audience in science?

Deehan: Audiences for mainstream programmes on BBC1 TV are between 10 and 14 million, with an average of 12 million. That size of audience is a straight representative slice of the UK population. The Horizon audience of about 3 or 3.5 million begins to be skewed towards the A, B and C1 (professional, employers and managers, intermediate and junior non-manual) sections of the population but it includes a large number of C2s (skilled manual workers). A typical radio audience may number about 1 million. The make-up of the audience for science programmes is about the same as for the network itself. In radio we can be sure that people tune in because they are interested in a programme, whereas many people watch television programmes because they happen to have the television on. The audience for our radio programmes tends to live in the south-east, be about 60% women, and come from the higher income and higher educational levels.

I don't know how much reliance we can place on these figures. When the numbers go below a million the statistical methods we use for gaining our audience information are right on the limits of significance.

Friedman: So far we have been talking about carefully produced documentary programmes. In the United States, if we want to talk about an audience approaching 70–80% of the population we have to look at the regular news coverage. Reporters have no science training and stories are condensed to 70 seconds or so, not 30 minutes.

Deehan: I was talking about science broadcasting in general. The minimum time we would spend on a topic in one of my programmes is about five minutes. But when I talked about the problem of scientists not getting their message over to the public I was thinking particularly of news and current affairs programmes, where a minute or so of air time is not uncommon.

In the USA there are thousands of radio stations and it is not surprising that they have no science specialists. Yet reporters from the boondocks turn up at AAAS meetings and have to try to interview scientists about oncogenes and so on.

Friedman: Scientists should get their metaphors together and try to be more humorous but we also need to look at preparation for the journalists. The Kemeny Commission report (Kemeny 1980) on the press coverage for the Three Mile Island disaster concluded that the best reporting was done by the youngest science writers, who were fairly up to date in their science, and by the toughest general reporters, who refused to be put off.

Deehan: There are two broad divisions in the reporting of science: straight science reporting and reporting political matters that have a scientific input. I don't have the skills necessary to grill a politician. It is not surprising that a trained political reporter could uncover the bureaucratic camouflage at Three

Mile Island. In reporting these matters to the public there are two aspects: the political explanation and the scientific. A reporter dealing with scientific matters needs some appreciation of science and curiosity about it but need not be a scientist.

Rhodes: Television has to use pictures and pictures often get in the way of argument. Television is a bad medium for conveying either argument or ideas but very good for conveying emotion. Radio is quite good at emotion too but a starving child on television arouses a lot more emotion than one on radio.

Another difference between radio and television is the size of the audience. Sometimes 40% of the total TV audience at a given time is watching one of our television programmes. The series on hospital life earlier this year reached 37 million people during the week. On the other hand, a lot of people see programmes just because they have turned the set on and flicked through the channels, as Geoff Deehan mentioned.

Television programmes of course cost an enormous amount to make. A 50-minute documentary costs over £100 000 to make in the UK.

Another point about news coverage and the appetite for scientific information is that when the Chernobyl and Challenger disasters happened, the audience for the evening news on BBC1 jumped from an average of 8.2–8.5 million to 10.5–10.7 million the first evening and went even higher the next evening. But although the reporting after Chernobyl was good, we had failed to prepare people beforehand; they had insufficient background knowledge to understand what was happening.

Laetsch: Later we might discuss what the implications of these enormous audiences are for scientific literacy. Can the people who watch or listen to television or audio programmes pass Jon Miller's test of scientific literacy?

Interactive radio might be useful. Some months ago a radio station in Iowa announced that people should protect their ears because the telephone lines were going to be cleared out. Several stores then sold out of plastic bags which people bought to put over their telephone receivers. That is a wonderful example of scientific illiteracy but it also shows that radio can get messages over. Another example is that we have school teachers in the USA who think that electric light bulbs blow out because the electricity in the bulb has gone.

Miller: Just after the Challenger explosion in January 1986 we re-interviewed people we had interviewed in November and December 1985 on space issues. Within 18 hours of the spacecraft explosion, 94% of the American people had seen pictures of the explosion. The interest in space among the group we interviewed went up after the disaster, but their self-reported sense of being well-informed went down; the more media they consumed, the less likely they were to say they were well informed, which we found confusing.

Rhodes: Until that moment they knew it could never happen. When it happened all their previous concepts fell apart.

Falk: The whole inquisition in the press was predicated on the questions:

'how could this happen?' and 'we don't understand how this happened'.

Rhodes: The same thing happened after Chernobyl. People felt they were woefully ill informed, though before the event they would have said 'I know all this'.

Tisdell: It is a case of the more you know, the more you know you don't know.

Laetsch: Educators call this kind of thing a discrepant event, and that is used as a basis for a lot of subsequent education.

Reference

Kemeny JG 1980 The accident at Three Mile Island. Pergamon, Elmsford, New York

Selling science: how the press covers science and technology*

DOROTHY NELKIN

The Program for Science, Technology & Society, 632 Clark Hall, Cornell University, Ithaca, NY 14853, USA

Abstract. The images of science and technology conveyed by the press are critical for the public understanding of science. Using examples mainly from biomedical science, I explore how scientists and their institutions influence press coverage through their use of public relations and communication controls. I suggest that scientists themselves help to shape the images of science and colour the information that reaches the public through the press. I also explore the tensions between the two cultures of science and journalism, analysing some fundamental differences that must be understood in the interests of improved communication.

1987 Communicating science to the public. Wiley, Chichester (Ciba Foundation Conference) p 100–113

It is fashionable in these days of great concern with science literacy to criticize the media for inaccurate or incompetent coverage of science. For example, George A. Keyworth II, President Reagan's former science adviser, accused the media of irresponsibility, of emphasizing only the hazards and not the benefits of emerging technologies. He believes that 'the press is trying to tear down America' (Keyworth 1985). Philip Handler, former president of the National Academy of Sciences, wrote in 1980 that 'anti-science attitudes perniciously infiltrate the news media'. A UCLA chemist accused the press of 'an intent to show how modern technology is poisoning America'.

The press coverage of science is easy to criticize. Exposed through their extensive writing to public scrutiny, journalists are an open target. I would rather focus on an aspect of science journalism that is seldom discussed. Where do journalists obtain their information, and in what form? How are the images of science and technology that are conveyed to the public influenced by the efforts of scientists themselves?

---

* The material in this paper has been abstracted from my book, *Selling Science: How the Press Covers Science and Technology* (W.H. Freeman, New York, 1987)

First I would like to emphasize the importance of the press for public knowledge about critical science-based issues. Most people understand science and technology less through direct experience than through the filter of journalism. Newspapers, popular magazines and television are their only contact with what is going on in technical fields. Journalists in effect are brokers, framing social reality and shaping the public consciousness about science. Through their selection of news they set the agenda for public policy. Through their disclosure of new discoveries they affect consumer behaviour. Through their style of presentation they lay the foundation for public attitudes and actions. Media coverage of science may have implications for the distribution of scarce resources; access to the media can bring in research funds. (For a perspective on the role of the press see Hall 1979. This 'agenda-setting' role of the press has been widely documented: see McCombs & Shaw 1972, Cohen 1963, Murdock 1974, Tuchman 1978.)

In this context, control over the information and images, the values and views, the signs and symbols conveyed to the public is understandably a sensitive issue. Industries, political institutions, professional groups and aspiring individuals all want to manage the messages that enter the cultural arena. Scientists today, often working on the border between advanced research and its application, are no exception. Indeed, the quest for publicity is increasingly prevalent in the scientific community, as many researchers believe that scholarly communication is no longer sufficient to maintain their costly enterprise—that national visibility through the mass media is strategically necessary to assure a favourable public image and therefore adequate support. Thus they are using increasingly sophisticated public relations techniques to control the language and content of science news. Let me provide some examples:

In 1985 Kenneth Wilson, Nobel Prize winner in physics, helped to convince the National Science Foundation to provide $200 million for the creation of major supercomputer centres in four universities. He attributes this successful quest for funds to a newspaper article that quoted some scientists to the effect that without a nationally funded programme the United States would lose its lead in supercomputer technology. 'The most amazing thing to me', said Wilson, 'was how the media picked it up and [how] a little insignificant group of words could change everything.' In fact, this group of words established a powerful image that reinforced Wilson's goals. Wilson, meanwhile, learnt a basic principle in public relations:

The substance of it all [the supercomputer research] is too complicated to get across—it's the *image* that is important. The image of this computer program as the key to our technological leadership is what drives the interplay between people like ourselves and the media and forces a reaction from Congressmen. (Wilson 1985.)

This principle increasingly guides the expanding public relations activities in major research institutions.

Public relations in science is hardly new. In 1847 Joseph Henry, the physicist and first secretary of the Smithsonian Institution, wrote of the need for public representation of science: 'It is evident that the principle means of diffusing knowledge must be the press' (in Krieghbaum 1941). Nineteenth-century scientists publicized their own work by writing for the press. But then, with the increase of private philanthropy and industrial support of science at the turn of the century, popularization declined and indeed began to appear unseemly as a professional activity (Tobey 1971).

Between the two world wars an expanding scientific enterprise needed public support, and once again scientists sought ways to enhance their public image. Professional associations began to organize public relations departments. As the scientific enterprise has grown in cost and complexity since World War II, public relations has been viewed as increasingly important as a means to enhance institutional prestige, to establish competitive advantage in 'hot' fields of research, to encourage public support, and to influence public policy.

Public relations efforts are growing in every field, but they are most elaborate in the promotion of dramatic medical interventions, promising new discoveries or therapeutic techniques. The press has been highly responsive in this area and, indeed, prone to exaggerated claims. Let me use some examples of biomedical reporting to illustrate some of the problems of science journalism and also the vulnerability of reporters to the claims of scientists and their institutions. Take, for example, the media coverage of organ transplants, which began in the 1950s with reports of a 'revolutionary new stage of medicine . . .' In 1958 *Life* magazine portrayed transplant techniques as the ultimate solution to the most fundamental problems of life and death. The reporter predicted 'when kidneys can be transplanted, two kidneys will be a luxury.' There is 'promise of a glowing future'.

By 1968 the glowing future arrived. Journalists hailed Dr Christiaan Barnard's dramatic operation as a 'surgical landmark' comparable to space exploration. Barely mentioned was the fact that his patient died. The press subsequently heralded heart transplants with front-page headlines. Primed by enthusiastic press releases from medical centres, reporters conveyed the misleading perception that heart transplant operations were 'miraculously effective' solutions for heart patients. They gave little attention to the long-term implications, i.e. the patient's postoperative histories or their deaths. The transplant was a dramatic event; the aftermath ceased to be news.

Optimistic images again flourished in the coverage of the first artificial heart experiment at the Utah Medical Center, which employed a public information staff to work with the hundreds of journalists who arrived in Salt Lake City to cover the event. News reports of the 'dazzling technical achievement', 'the blazing of a new path', the 'medical milestone', continued for 112 days—until the death of Barney Clark. While the Center provided a deluge of technical detail, critical information that would have helped reporters to understand the policy issues remained undisclosed. How were decisions made concerning the selecting of the patient and the choice of the Jarvik-7 model of the artificial heart? What did the scientists learn? What was the role of the Institutional Review Board and the nature of its deliberations? Without such details, reporters, and thus readers, were only inadequately informed (see Altman 1984).

Occasional scepticism crept in as Utah reporters, many of whom were Mormons, questioned the metaphysical implications of replacing a heart by a machine: 'The heart is the symbol of love, site of life, habitat of the soul. Can it be replaced by a simple mechanical pump?' Elsewhere the ethical issues emerged in cartoons. The *New Yorker* portrayed some doctors observing a patient: 'Our hope now is to just get him strong enough to pull his own plug.'

When the artificial heart team moved to Humana Hospital in Louisville, Kentucky, this medical centre promptly awarded a contract to a professional public relations firm to orchestrate the media coverage of its artificial heart experiment on William Schroeder in December 1984. In a special media centre physicians and public relations staff briefed reporters daily on the technical as well as the personal aspects of this medical intervention. However, the briefings oscillated with Schroeder's state of health. In the words of one reporter, the public relations staff shifted from a NASA-style to a Soviet-style approach as the hospital carefully managed the material it wanted to reveal to the press.

The result? We read more about the stories of success than the process, the dead ends and wrong turns. Informed of individual accomplishments and spectacular events, journalists convey little about the sociology of science, the structure of the research institutions, or the detailed, long-term and incremental process of research. While social or ethical problems occasionally make their way into print, little appears on the values that guide decisions about the costly and controversial procedures. Science becomes more a subject for consumption than for critical public scrutiny, more a source of entertainment than a matter of social choice. And these tendencies, I am arguing, are strongly encouraged by those seeking to sell science, namely scientists and their institutions.

In less dramatic scientific and technical areas, public relations efforts serve pragmatic purposes. Sometimes new scientific discoveries and therapeutic techniques are promoted as a means to win institutional prestige. For example, in October 1984 a research group at a Dartmouth, New Hampshire, medical centre held a press conference to announce the results of the clinical trial of a new therapy for Alzheimer's disease. In a published technical paper the scientists had explained that their research was very preliminary, having been tried out on only four patients. But their press release failed to mention the study's limitations, and the decision to hold a press conference turned the research into a media event. Not surprisingly, the media headlined the research as a 'breakthrough' and a 'successful treatment'.

Science also appears in the press when firms try to market products by defining them as newsworthy scientific discoveries. Journalists are often vulnerable to such promotional techniques. For example, in the 1960s and 1970s physicians and drug firms strongly promoted oestrogen replacement therapy, claiming it would reduce the biological effect of ageing. The major source of information for journalists was the director of a foundation funded by three drug firms to distribute reports about specific products. He and several other physicians made sure that promotional materials on oestrogen were well designed to attract the press, which was, of course, attracted in any case to stories about a therapy that promised eternal youth (see Kistner 1969).

In the mid-1960s news articles on oestrogen replacement therapy suggested that the drug was a means of reducing the biological effects of ageing. The headlines of these articles read: 'Science Paints Bright Picture for Older Women'. An Associated Press newswriter cited a scientist: 'Women are fortunate people: there is no reason why they should grow old.'

The discovery of a pill which would keep women young is surely a newsworthy event. Eager for copy, journalists uncritically accepted the claims of interested experts who debunked the growing set of studies that suggested a relationship between oestrogen and endometrial cancer. These claims continued even after the Food and Drug Administration issued warnings to that effect.

When US Senate hearings in 1970 on the safety of oral contraceptives called public attention to the evidence of side-effects, some reporters conveyed doubts about using oestrogen for birth control, but their articles on oestrogen replacement therapy also cited 'reassuring truths', and 'reassuring polls'. They uncritically cited oestrogen proponents who dismissed the concern about risk by observing 'When we drive down the freeways, we take a risk.'

Drug companies frequently use science-based 'press agentry' to market products, in effect pushing them as newsworthy discoveries. Lilly's arthritis drug, Oraflex, was initially marketed this way. In 1982 the firm's public relations office sent out 6500 press kits promoting the drug by using scientific evidence to support the claims of its effectiveness in relieving arthritis. Lilly dispatched scientists around the country to contact smaller newspapers. Some experienced science reporters refused to cover the story, suspecting that Lilly's claims were exaggerated. However, the product was covered as science news in 150 newspapers and television stations, and prescriptions increased from 2000 to 55 000 a week. Then a report showed its harmful side-effects and, after only 12 weeks, it was withdrawn from the market.

Physicians have used similar marketing techniques to promote the use of cortisol antagonists in the treatment of anorexia. This controversial therapy, designed to reduce the level of the hormone cortisol in the brain, was based on a study of 33 patients. Rather than submitting the findings to a scientific journal, the physicians announced the results at a press conference organized by a public relations firm. 'One can't afford to take the time it takes through the medical journals', said one of the doctors. The rush was related to their effort to market a proprietary line of nutritional products and to expand a private anorexia clinic.

There are many examples of how science-based press agentry is employed to build public confidence in other technologies. In the 1970s the nuclear industries worked closely with the press to project positive images in order to allay public fear of nuclear power. Industry PR officers proposed substituting palatable synonyms for scare words such as 'hazard' or 'criticality'. Soon nuclear plant sites became 'nuclear parks' and accidents became 'normal aberrations'.

In 1975 consultants for the electric power industry outlined a 'nuclear acceptance campaign' involving scientists who travelled around the country to lecture and to make themselves available to reporters. Public relations firms, specialized in running political campaigns, trained the scientists for public debate and taught them how to approach the media. The idea was that 'The public has faith in science, believes scientists and would listen.'

The chemical industry uses similar strategies. There is a so-called 'visible scientists' programme in which scientists trained by public relations firms travel around the country packaging industry-generated material for local papers and television stations. Assuming that scientists have greater credibility than either businessmen or politicians, several public relations firms have tried to generate business by proposing that corporations develop 'parachute teams' or 'truth squads' of scientists, ready to move into risk situations in order to diffuse opposition.

Public relations professionals are in fact an important arm of the media. They often save editors and reporters hours of work tracing down the news. They contribute in very important ways to informing the public about products, ideas and services. Packaging complex material in manageable form, they serve as brokers or liaisons between science and the press. However, they are working for clients and their job is to make their institutions look good (Stainton 1977).

From the earliest days of public relations journalists have been ambivalent about its role. While they regard such efforts as a means to subordinate journalism to private interests, they are profoundly influenced by it—a fact which once prompted Upton Sinclair to define journalism as 'a business in the practice of presenting the news of the day in the interest of economic privilege.'

Today reporters and editors are more sceptical than ever. They complain of constant pressures from public relations professionals: 'I get calls from doctor knowledge, the world's leading authority on X disease or Y technology, who is also president of Z society.' They refer to 'pesky PR types', or 'the flacks'. Irritation is not limited to corporate public relations. 'They're all grinding the same axe, from breakthrough university to wonder pharmaceuticals to the National Institute of Nearly Cured Diseases.' (These and other statements appear again and again in letters and discussion in the newsletter of the National Association of Science Writers: see Bloom 1979 and others.)

Some editors feel that their newspapers are used as pawns for grantsmanship. 'When NSF money was available easily, you couldn't get a story out of a molecular biologist. Today, I get copies of grant applications in the mail with this thing, "single cure for blank", or whatever the hell it might be, circled in red, saying "we need all the help we can get, fellers".' However, reporters do tend to trust scientists as sources more than politicians. Moreover, journalists must deal with technical complexity and scientific claims that are difficult to check out. Socialized to regard scientists as a reliable and objective source of information, and anxious to maintain future access to their sources, they are inclined to rely uncritically on material that is conveniently packaged by scientific institutions and their public relations staffs. Their vulnerability to expertise converges with the deadlines and competitive constraints of the news-gathering business to give an unusual degree of power to those sources who are best organized to provide facts in a manageable and efficient form. These sources of technical information have become increasingly sophisticated about the needs of the press, thereby enhancing their control over the news.

Despite the increase in public relations, many scientists remain ambivalent about the press. While public communication is seen as necessary and desirable, it also extends accountability beyond the professional community. For once information enters the arena of public discourse it becomes a public resource. Thus they seek various ways to control the images that appear in the press.

Professional journals, for example, are publishing guidelines on how to deal with reporters. They offer these guidelines as defensive tactics. An article in the *New England Journal of Medicine (NEJM)* suggests that scientists use the public relations office of the universities as a clearinghouse. It warns scientists to be aware of reporters' motives. 'Your response to an innocent sounding question about your study of schizophrenia may be linked in tomorrow's newspaper to a murder trial.' The *NEJM* warns the biomedical community to avoid interviews prior to publication: 'Never even whisper to a reporter anything you would not care to see in screaming headlines.' It suggests that research investigators do a dry run with a public relations officer before an interview, that they elicit a promise that the reporters will show them articles before publication, and that they tape the interview so as to have an exact record. 'If you feel trapped, obfuscate: it will get cut if it is too technical' (Bander 1983; see also Miller 1978).

Scientists control science news in part by discouraging their colleagues from 'going public'. In her book *The Visible Scientists* (1977), Rae Goodell observes that scientists who become visible 'are typically outsiders, sometimes even outcasts among established scientists . . . seen by their colleagues almost as a pollution in the scientific community—[They] are breaking old rules of protocol in the scientific profession, questioning the old ethic, defying the old standards of conduct.'

The inclination to avoid reporters is encouraged by the policies of a number of professional journals: the *NEJM*, the *Archives of General Psychiatry*, *Science* and several other journals will not consider an article whose content has been published elsewhere, especially in the popular press. These policies have been a source of angry discussion focusing on the 'Ingelfinger rule' (after the late editor of the *NEJM*) that guides the publication policies of the *NEJM* (Ingelfinger 1970).

Ingelfinger's successor, Arnold Relman, has perpetuated the rule, emphasizing his responsibility to maintain the reliability of scientific information through the system of peer review. He argues that the public interest is not well served by disregarding this system, for journalists could raise hopes or fears on the basis of false or unreliable information (e.g. Relman 1981).

Journalists, however, are appalled by the Ingelfinger rule, arguing that it violates the public's right to know and reinforces the cautious behaviour of scientists, who are often reluctant in any case to talk to science writers. The journalists cite areas in which delayed publication has caused public harm. For example, a newspaper story about early laboratory research on the effects of 'smoking' on beagles delayed publication of the findings of this cancer research because scientific journals refused to publish the results on the grounds of prior disclosure to the press. Indeed, due to such concerns *JAMA*, the *Journal of the American Medical Association*, has refused to go along with the rule. Relman himself has made exceptions in the *NEJM*; early findings of research on toxic shock syndrome and AIDS, for example, can be disclosed because early public knowledge of medical information on these issues is urgent for public health.

Scientists also try to control press coverage by refusing interviews unless they can review the copy before publication. Reporters, fearing censorship by vested interests, are reluctant to show their articles to sources, though they often check details for accuracy. As science writer Victor Cohn explains, 'Scientists are to reporters what rats are to scientists. Would scientists ever allow their subjects to check the interpretation of their behavior?'

Communication controls reflect the persistent tensions between scientists and journalists, tensions which reflect some fundamental differences between the two professions. Let me conclude by laying out these differences, for I believe it is in the interest of improved public information to understand them.

To begin with, scientists and journalists often differ in their judgements about *what is news*. In the scientific community, research results become reliable and therefore newsworthy through the endorsement of professional colleagues. Research findings are provisional—and therefore not newsworthy—until certified by peers to fit into the existing framework of knowledge.

For journalists, however, proven and established ideas may be of less interest than new and dramatic, though possibly tentative, research. Thus they are attracted to non-routine, non-conventional and often aberrant events. Yet journalists too are troubled when over-zealous researchers seek press coverage of their work before the time-consuming process of peer review. A recent case in point was the French announcement of the 'dramatic' results of using cyclosporin for AIDS patients. Journalists who suspected the press release could not ignore it for fear of being scooped. They published the report, but then scientists admitted the disclosure was premature.

A second source of tension lies in the conflict between the professional practices of journalism and scientific expectations about *appropriate styles of communication*. Constrained by deadlines, limited space and the interests of their readers, journalists must select and simplify technical information. This often precludes the precautionary qualifications that scientists feel are necessary to accurately present their work. Readability in the eyes of the journalist may be over-simplification to the scientist. Indeed, many accusations of inaccuracy follow less from actual errors than from efforts to present complex material in a readable and appealing style.

Differences in the use of *language* contribute to strain. The language of science is precise and instrumental. Information is communicated for a purpose—to indicate regularities and aggregate patterns, and to provide technical data. In contrast, journalistic language is often chosen for richness of reference and suggestiveness. Scientists direct their professional communication to an audience trained in their discipline. They can take for granted that their readers share certain assumptions and therefore will assimilate the information in predictable ways. When scientists write reports they often forget that some words may have special meanings in a scientific context and may be interpreted differently by the lay reader. For example, confusion over the definition of 'evidence' is a frequent source of misunderstanding. Biostatisticians use the word 'evidence' as a statistical concept. For biomedical researchers, the critical experiment is also defined as evidence. Most lay people, including journalists, accept as evidence anecdotal information or individual cases. While scientists talk of aggregate data, reporters write of the immediate concerns of the individual reader.

However, the most important source of strain between scientists and journalists lies in the ambiguity about the appropriate role for the press. Scientists often talk about the press as a conduit or pipeline, responsible for converting science into a form where it may be easily transported to the public. Confusing their special interests with general questions about the responsibility of the press, they are reluctant to tolerate coverage of the limits or flaws of science. Regarding the press as a technique to further scientific goals, they expect to control the flow of information to the public just as they do within their own domain. And they feel betrayed when their views are challenged. Even the most tempered and factual reporting can provoke a defensive response. When Harold Schmeck, science reporter for the *New York Times*, wrote a non-sensational article about interferon research in which he warned his readers not to hope for immediate miracles, a group of scientists complained in a letter to the editor that such expressions of doubt could affect their research funding.

Many science writers, living between the cultures of science and journalism, are themselves ambivalent about their role. Journalists traditionally view themselves as an independent profession, a kind of watchdog engaged in probing, not promoting, the institutions they write about. But in science writing ambivalence is reflected in a minimal amount of probing investigation, bold interpretation and critical inquiry. While the press today publishes criticism of art, theatre, music and literature, science is usually spared. While political writers aim to analyse and criticize, science and medical writers seek to elucidate and explain. While political reporters go well beyond press briefings to probe the stories behind the news, science writers tend to rely on scientific authorities, press conferences and professional journals to uncover the news. Many journalists are, in effect, retailing science and technology more than investigating them, identifying with their sources more than challenging them.

If the popular press is to mediate between science and the public, to help the public make complex judgements about crucial policy issues, and to enhance the credibility of scientific and technological choices, both journalists and the scientific community must come to terms with a different mode of communication. It is not enough merely to react to technical events, translating and elucidating them for popular consumption. Improved communication about science and technology calls instead for searching investigation and critical interpretation. Developing this kind of probing journalism will require greater tolerance on the part of editors, who must be willing to provide opportunities and rewards for investigative reporting on science and technology. It will require a well-trained cadre of journalists who are inclined to view science and technology as a serious and challenging beat. And it will require a willingness among scientists to distinguish critical inquiry from anti-science criticism, to restrain the enthusiasm that contributes to oversell, and to encourage the probing of competent journalists in the interest of public understanding of a central social institution.

References

Altman LK 1984 After Barney Clark: reflections of a reporter on unresolved issues. In: Shaw MW (ed) After Barney Clark. University of Texas Press, Austin, p 113–128

Bander M 1983 The scientist and the news media. New England Journal of Medicine 308:1170–1173

Bloom M 1979 The editor's letter. NASW Newsletter (April)

Cohen BC 1963 The press and foreign policy. Princeton University Press, Princeton, New Jersey

Goodell R 1977 The visible scientists. Little Brown, Boston, p 61

Hall S 1979 Culture, the media and the ideological effect. In: Curran J et al (eds) Mass communication and society. Sage, Beverly Hills, chapter 13

Handler P 1980 Public doubts about science. Science 208:1093

Ingelfinger F 1970 Medical literature: the campus without tumult. Science 169:831–837

Keyworth GA II 1985 [Interview with F. Jerome.] SIPIScope (February)

Kistner R 1969 The Pill: fact and fallacies. Delacorte Press, New York [see also articles in the Ladies' Home Journal and other women's magazines in 1969]

Krieghbaum H 1941 American newspaper reporting of news. Kansas State Bulletin, p 20 (15 August) [quoting Joseph Henry]

Life 1958 (14 July)

McCombs ME, Shaw DL 1972 The agenda setting function of the mass media. Public Opinion Quarterly 36:176ff.

Miller NE 1978 The scientist's responsibility for public information: a guide to effective communication with the media. Society for Neuroscience, Bethesda, MD

Murdock G 1974 Mass communication and the construction of meaning. In: Armstead N (ed) Reconstructing social psychology. Penguin, Harmondsworth, Middlesex

Relman AS 1981 The patient and the press. Bryn Mawr Alumni Bulletin, p 2–5 (Fall)

Stanton I 1977 Public relations and the press. In: Wagner G (ed) Publicity forum. Weiner, New York, p 31

Tobey R 1971 The American ideology of national science (1919–1936). University of Pittsburgh Press, Pittsburgh

Tuchman G 1978 Making news. Free Press, New York

Wilson K 1985 [Quoted in] New York Times (16 March)

DISCUSSION

Dixon: Public relations activities are often disparaged but PR is a legitimate activity. What matters is what journalists do with the material that comes from PR companies. At one stage when I was editor of *New Scientist* the journalists working for its parent publishing company had to work to rule for a time. The instructions issued by the National Union of Journalists included one that said 'Press releases will no longer be simply marked up as usual and sent to the printer'. Clearly this kind of mark-up was accepted by the NUJ as a normal activity. On some publications—though not, I emphasize, *New Scientist*—press releases are usually sent off to be printed as they arrive, with a few quote marks around them.

When the *Observer* newspaper printed a story by the medical correspondent some years ago saying that the National Heart Hospital in London was preparing to do heart transplants the British Medical Association issued a press release on behalf of the hospital saying categorically that there were no plans to do this. Yet plans were being made, a team was being brought together and six or seven months later the National Heart Hospital did its first heart transplant. In between, Dr Christiaan Barnard had done his first operation. Again what matters is what the journalists did with the press release.

Nelkin: One study of environmental reporting defined the role of the journalist as simply opening the mail.

Friedman: You mentioned that some scientists want to review stories before they appear. There are two possible requests. One is to say 'You cannot use the story unless I approve it as written'. The other is to say 'I would like an opportunity to comment after seeing your story. I might be able to clarify some points or correct others, though of course you don't have to use my comments'.

Nelkin: Science journalists will often call back after an interview and check the accuracy of pieces of information. However, in the USA journalists do not normally show a whole story to an interviewee. They fear political or interest group censorship and they have time deadlines. Journalists have neither the time nor the money to send articles back. But they will on request send copies of the piece after it has been published. As a member of the Media Resource Service I am interviewed often. I never ask to see a copy before publication. If I insisted, the journalist would dump the article.

Rhodes: We would do the same in television documentary making. If a scientist demanded to see a near-final version and have the right to change things, we would simply pull out before we filmed the interview.

Friedman: There is a difference between insisting on the right to change things and requesting the opportunity to point out to journalists that they said something was black when it was white.

Rhodes: I agree but it is a question of whether time allows for the story to be read over the phone.

Nelkin: Calling back to check technical points is not the same as showing the whole article, which may contain other material that the scientist would not agree with.

Friedman: But we are talking about the generalized statement at the end of the facts. In New York recently a multi-billion dollar project was cancelled partly because of the word 'significant'. The word was quoted accurately but the question was its interpretation in the next paragraph. The scientist intended 'significant effect' to mean just above the statistically insignificant level. The interpretation by the politicians and the press was that the word meant a large effect.

Nelkin: Journalists should learn more science but scientists should learn to use more non-technical language. If scientists want to get their ideas across they have to understand that they are not talking to other scientists. The words 'significant' or 'evident' or 'epidemic' might be interpreted in different ways by the lay public than by scientists.

Thomas: You described journalists as being vulnerable to various pressures, in particular the pressures of a deadline and their own competitive instincts. But most members of the public in the UK read only one newspaper and probably wouldn't be aware that a story had come out in another newspaper the day before or the week before. So why the anxiety by journalists to get a story out on a particular day?

Nelkin: Some of the journalists here could explain that better than I can. I think science journalists are vulnerable to their sources for several converging reasons, not just competition. Science journalists are often in awe of science. They go into science journalism because they like science and many of them wish they were scientists themselves. Like sports writers they stay in their profession and don't shift around to different fields. They have deadlines and feel competitive pressures but they are also dealing with areas of great complexity. They don't quite trust themselves, and they don't usually have time to do research (though Lawrence Altman was given six months to do a series on AIDS in Africa). They all go to the same press conferences and worry that 'If the *New York Times* covers it and I don't, my editor is going to get on my back'. The competition comes largely from editors who put a tremendous amount of pressure on journalists to cover the hot stories.

Tisdell: Another possible obstacle to science journalists and co-operating scientists is that the editor of a publication might not accept 'their' article. If it is accepted, the editor might cut a lot out of it.

Nelkin: Science journalists often have more autonomy than other kinds of journalists, because some editors don't know as much about the subject.

Whitley: A lot of what is communicated in science is written for administrators of science or government agencies. This should receive more attention. We published some articles on it last year in a book on popularization (Shinn & Whitley 1985). Not much sociological work has been done on the ways in which some scientists use the media to communicate to other people. They popularize science in the sense of producing information for an audience of non-specialists. The growth of specialization in science and the need to compete for funds through a peer review system means that proposals have to go to people who are not necessarily specialists in the same field. This reflects a change in the nature of scientific research, particularly compared to the 19th century when much of the funding was done by the home institution. This is more significant than worrying about whether scientists can talk to journalists in the same language.

Nelkin: I referred before to both pork-barrel funding and corporate funding. People in Congress and corporate executives read the press. Scientists thus have a stake in influencing the media.

Laetsch: One of the differences between pork-barrel funding and what you were talking about is peer review.

Nelkin: It is usually not peer reviewed.

Whitley: Some proposals go through both the formal system and the public system. In the USA it is clear that cancer researchers try political pressure, public pressure and public relations pressure as well as getting influential people in the scientific and medical elite on their side.

Nelkin: Many scientists feel that they must enhance their public image; that doing good work is no longer sufficient to obtain substantial funds. Thus, they are sensitive to press coverage.

Reference

Shinn T, Whitley R (eds) 1985 Expository science: forms and functions of popularisation. Reidel, Dordrecht (Sociology of the Sciences Yearbook 9)

Museums and the communication of science

Department of Public Services, British Museum (Natural History), London, SW7 5BD, UK

Abstract. Statistics for the number of museums and numbers of visitors to museums are, in isolation, impressive, and it has been claimed that museums are important sources for informal learning of science. There are reasons to doubt this claim for traditional Western museums. In reality museums reach relatively few people, from a restricted sector of society; and evaluation shows that little learning goes on among 'casual' visitors. Data from large-scale sample surveys, observation studies and research into visitors' perceptions of exhibits are generating a natural history of visitor behaviour. Various theories have been proposed to account for the behaviour of science audiences, in museums and in other situations. One of the variables in Grunig's theory, level of involvement, determines whether people process science information passively, for pleasure and to pass the time, or actively, to solve specific problems. The former behaviour is associated with the mass media and, it is now clear, museums; the latter is associated with specialized sources. This has implications for the educational goals museums should adopt, and how they should set about achieving them.

1987 Communicating science to the public. Wiley, Chichester (Ciba Foundation Conference) p 114–130

The Royal Society's report on *The public understanding of science* (Bodmer et al 1985) asserts that 'Museums are a major informal mechanism for effecting public understanding of science' and similar statements can be found elsewhere in the literature. I propose to start, therefore, by looking at the scale of museum operations; I shall then consider how museums are used by the general public and what role museums might realistically aim to fulfil in the communication of science. The main area of concern is the widely held yet erroneous belief that science can be communicated to a mass audience simply by delivering the information in an efficient way (Grunig & Disbrow 1977).

The museums

There are more than 35 000 museums in the world (Hudson & Nicholls 1985) if, for the moment, we include science centres and living displays. These

TABLE 1 Museums and population data for some advanced countries (data from Hudson & Nicholls 1985)

Country	Number of museums	Size of population (millions)	Museums to population index (MPI)[a]
Finland	606	4.8	12.6
Australia	706	7.6	9.3
New Zealand	255	3.2	8.0
Canada	1514	24.6	6.1
German Democratic Republic	748	16.7	4.5
German Federal Republic	2415	61.7	3.9
Great Britain	2127	55.9	3.8
France	1921	54.1	3.6
United States	7892	222.5	3.5

[a] Number of museums for each 100 000 of the population.

museums are unevenly distributed, but data for some developed countries are given in Table 1. Finland has the highest museums-to-population index (MPI, number of museums for each 100 000 of the population) in the world, almost four times that of the United States. Not surprisingly, in view of their more urgent priorities, developing countries tend to have low MPIs. However, among others, India (MPI 0.07) has experienced a marked growth in the development of science museums over the last 20 years, with the express purpose of improving out-of-school science education (Danilov 1982).

The proportion of the world's museums that deal in part or in whole with science is unknown. However, about one-fifth of all museums are found in the United States, so we can profitably look at some statistics for that country. Kimche's (1980) figures for 1979 give an estimate of 5500 non-profit museums (cf. Table 1), of which 18% are science museums (including aquariums, zoos and so on). In 1985 these non-profit science museums received 45% of all visitors to museums, i.e. 158 million out of a total of 353 million. In the same year professional baseball, basketball and football matches received a total of 80 million visits. By way of comparison, there were 49 million visits to museums and 24 million to professional football matches in Britain in 1982 (Kemp 1986).

We should not be misled into exaggerating the educational opportunity these figures seem to promise. The estimated average attendance for science museums in the United States is 190 000 visitors a year (the most popular, the Air and Space Building of the Smithsonian Institution, receives an exceptional 10 million a year). In Britain, the Science Museum and British Museum (Natural History) each claim in the region of 3 million visitors a year. It is salutary to compare these figures with those for television science. In the United States, the average audience for single *Nova* programmes is 10 million (La Follette 1982), and in Britain *Horizon* programmes average 3.5 million. If we extend this comparison to newspapers, the current circulation figures for serious papers like *The Times* and *The Guardian* in Britain are close to half a million a day, while the corresponding figure for the downmarket *Sun* is four million. While it would be unprofitable to debate what is and is not a mass medium, we should hesitate before claiming museums as significant sources of information for the public at large, or before comparing them with the popular mass media.

The visitors

Although many museums have carried out visitor surveys, these have been uncoordinated and often done for no good reason, so that on the whole we are still ignorant about who goes to science museums, why they go there, and what they get out of the experience. This has led to *non-conflicting* pleas from Loomis (1973)—*Please! not another visitor survey*—and from Laetsch (1979), for the construction of a natural history of visitor behaviour. In the absence of good general information, I am forced to concentrate on London's Natural History Museum, the major division of the British Museum (Natural History). However, I believe the overall picture that I present holds true for most, if not all, large traditional science museums in the West. My sources are Alt (1980), Griggs & Alt (1982) and unpublished reports.

Ten per cent of the Natural History Museum's visitors are children in organized school parties. I shall ignore these in this paper and concentrate instead on the non-coercive use of the Museum by 'casual' visitors. Roughly 25% of all visitors are overseas tourists; 75% are over 16 years of age; and about half come as members of family groups, fewer than 50% of which include children. In contrast to the UK population (48% male), visitors include slightly more males than females (54:46). In socioeconomic terms the Museum has a high class profile: three-quarters of its visitors are ABC1 (professional, managerial and skilled non-manual workers), with about 40% in the AB class. More than half (about 60%) have completed their full-time education, but 80% have no formal qualifications in biology, and less than 5% have a university degree in a biological subject. Nearly half of all visits to the Museum are planned on the day of the visit, and about 17% on the day before. The overwhelming proportion of visitors (about three-quarters) have not visited the Museum before in the previous 12 months, while only about 10% have visited twice or more in this period. Around 60% of all visitors spend between one and two hours in the Museum (median length of stay, 1 hour 47 minutes) with only about 20% staying for longer than two hours. The main reason that over two-thirds of visitors give for visiting the Museum is general interest and curiosity, including accompanying someone else. Only very few (about 2%) come to seek specific information in connection with their work or studies, although rather more (about 13%) come to see specific permanent exhibitions. Taken as a whole, the numbers in this paragraph contradict the received view of the Natural History Museum as a children's museum (Griggs & Hays-Jackson 1983). But they do show that for most people a visit to the Museum is a social activity.

Laetsch (1979) has recorded that visitors to the Lawrence Hall of Science in Berkeley, California, spend a high proportion of their time not attending to the exhibits; instead they are either in the cloakrooms, shops and cafeteria, or engaged in group management. The same pattern can be observed in the Natural History Museum and, as in other museums (e.g. Borun 1977), the mean time spent at individual exhibits is around 30 seconds. It seems that visitors have a limited amount of time to spend, and they move rapidly through the museum looking for interesting exhibits on which to spend it before 'museum fatigue' sets in, rarely spending much time on any one item for fear of missing something of greater interest just around the corner. Thus in the United States it has been found that the exit to a gallery tends to be an overwhelming attraction; and people are only really receptive to exhibits in the first half-hour or so of their visit (Melton 1935, Falk et al 1985).

Visitors' perceptions

Research by Alt (Alt & Shaw 1984) and S.A. Griggs (unpublished work 1984) has shown how visitors to the Natural History Museum perceive its exhibits. The results complement those from large-scale sample surveys and observation studies, rounding out our natural history of visitor behaviour.

Alt studied exhibits in the Hall of Human Biology at the National History Museum. First he interviewed visitors to elicit the characteristics *they* use to describe exhibits. Then a second group of visitors were asked to judge specific exhibits in the light of these characteristics, and also to say which of the characteristics would be present in their ideal exhibit. From this Alt derived the abstract concept of an ideal exhibit, which provided a reference point against which to compare both characteristics and real exhibits in a biplot analysis. Some of Alt's results are summarized in Table 2, which relates exhibit characteristics to the ideal. Griggs extended Alt's technique to make a comparative study of seven exhibitions in the Natural History Museum— three modern, three traditional and one a recently completed display with traditional characteristics. He also got visitors to rank characteristics according to their importance. A number of generally desirable characteristics were found to apply to all galleries, e.g. visitors expected the subject matter to be

TABLE 2   Relationship of exhibit characteristics to the ideal, from research in the Hall of Human Biology, British Museum (Natural History) (data from Alt & Shaw 1984)

Strongly negative	Orthogonal (neither strongly negative nor strongly apply)	Strongly apply
It is badly placed; not easily noticed It does not give enough information One's attention is distracted by other displays It is confusing	It is participatory It deals with the subject better than textbooks It is artistic It makes a difficult subject easier	It makes the subject come to life It makes its point quickly It has something for all ages It is memorable

well explained and easy to understand regardless of the style of exhibition design. However, a small number of characteristics were found to discriminate between galleries (Table 3). Griggs also studied these characteristics in relation to subgroups of visitors ('modernists', 'traditionalists' and those with no preference), but it will suffice to note that the lists of desirable and undesirable characteristics in Table 3 give us further insight into overall perceptions and expectations of the Museum's exhibitions.

Theories of visitor behaviour

Assuming the above data on visitor behaviour are typical, it is not surprising that, with rare exceptions, 'casual' visitors learn almost nothing from science exhibitions (Shettel et al 1968, Screven 1974). Clearly there is not much point in looking for large increases in factual information after a typical visit (Linn

TABLE 3   Desirable and undesirable characteristics of exhibitions listed in descending order of importance, as perceived by visitors to the British Museum (Natural History) (data from S.A. Griggs, unpublished work 1984)

Undesirable characteristics	Desirable characteristics
Subject matter not sufficiently explained Exhibits not realistic enough: difficult to relate to the real world It is appealing to children but less so to adults It is traditional in style; old-fashioned	It is obvious where one should begin and how one should continue It uses a lot of modern display techniques which help one learn It uses familiar things and experiences to make its points It includes a comprehensive display of objects and/or specimens

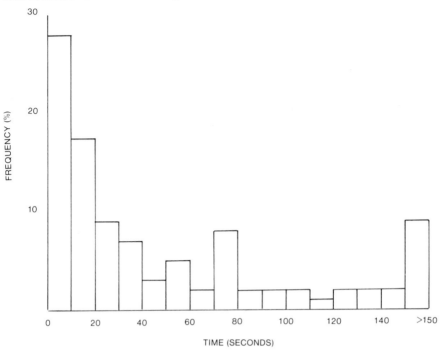

FIG. 1   Time spent by 'casual' visitors at an exhibit on hormones in the British Museum (Natural History). Data from Falk (1983).

1976). However, Borun's (1977) discovery that visitors to the Franklin Institute in Philadelphia showed a marked drop in positive feeling about the museum (from pre-visit to post-visit tests) shows it would be unwise to take refuge in the thought that all would be well if we could only find the right evaluation instruments *to prove* that visits to museums are valuable educational experiences. My own feeling is that it would be more useful to consider what could be achieved in educational terms, and under what conditions, by producing a theory that accounts for the behaviour and perceptions of visitors.

Falk's metaphor

One approach to the problem of visitor behaviour is given by Falk's (1983) analogy between shopping at a department store and visiting a museum. Based on bimodal frequency distributions obtained from studies of the time visitors spend at exhibits (Fig. 1), the analogy posits that most visitors to a museum are 'window shoppers', just wandering around to see what there is to see, with no more serious purpose in mind. A minority, however, are 'serious shoppers', who come intent on 'buying' messages from the few exhibits on which they spend their time. Window shoppers may, of course, submit to 'impulse buying' and thus become serious shoppers at the exhibits of their choice. However, window shoppers and serious shoppers are generally seen by Falk as two distinct groups.

There would seem to be two problems with this analogy. First, the number of serious shoppers predicted from a typical bimodal frequency distribution (9% in Fig. 1) is in excess of the 2% of visitors who come to the Museum in search of specific information. The analogy thus seems to put undue emphasis on the power of exhibits to turn visitors into impulse buyers. Secondly, then, it seems that this aspect of visitor behaviour (attending or not attending to exhibits) is better explained by M. Mann's argument (quoted in Grunig 1980) against the division of science audiences into 'readers' and 'non-readers'. Mann suggests that the groups readers and non-readers are not fixed; each group is volatile and changes with the circumstances, be these 'the story, the time, politics, weather . . .' In museum terms (excluding the small minority of visitors who come in search of specific information), different exhibits draw different audiences from among the same visiting population. This seems intuitively to provide an improved interpretation of Falk's data: one that looks at different behaviours within individuals rather than at differences between groups of visitors.

Grunig's theory of communication behaviour

While Falk is looking primarily at visitors' intrinsic motivations, Grunig (1980) provides an explanation of the behaviour of science audiences that would take the specific situations at exhibits into account rather than the motivations of visitors. It states, in effect, that how a visitor perceives a situation determines whether, and to what extent, he or she attends to the communication. The component concept of most interest here is *level of involvement*. This is the degree to which a visitor feels an involvement with the problem presented by the exhibit; in other words, how strongly the content of the exhibit relates to the life of the visitor. By way of illustration we may note that the desirable characteristic 'uses familiar things' and the undesirable characteristic 'not realistic enough' (Table 3) are readily related to level of involvement.

Level of involvement is of interest because it relates to whether visitors are processing information in a *passive* way, or are *actively* seeking information, i.e. in which of Mann's modes they are operating. Grunig (1979, 1980) remarks that people process information passively when they need to kill time and for enjoyment out of a sense of curiosity. They process this information by watching television, reading newspapers, looking at popular magazines and so on. On the other hand, people seek information actively when they have a particular problem to solve, i.e. when the information is going to be useful; and they seek it from specialized sources such as textbooks, hobby magazines, seminars and personal contact with knowledgeable people.

If crabgrass, for example, invades a lawn, the person will search for information on how to control it. But few people perceive an involvement with such scientific problems as black holes, whale populations, or animal genetics. If they have time available, people will process such low-involvement information when it comes to them randomly without any effort on their part, but they will internalize little of that information unless they are curious about the scientific problem—that is, recognize it as a problem. Those with high problem recognition . . . will seek out information related to the scientific problem and will remember the information they process. (Grunig 1980)

It requires a small leap of the imagination to see how well these conclusions fit with the results of various sorts of research into museums. Because it is not economical to do so, scientists and students of science rarely seek information in their own field from the mass media, including museums, despite the fact that exhibits in major museums have traditionally been designed for this purpose. Furthermore, the actual audience (Miles 1986a) in these museums is concerned with the consummatory rather than the educational use of knowledge, i.e. they are using it for pleasure and for passing the time in a largely passive way. Typically, there is little in the exhibition content of such museums to persuade them to do otherwise.

Conclusions

What lessons can be drawn from the above arguments? There are, I think, two main lessons that relate to the role of museums in the communication of science (Lewis 1980), assuming they have higher aims than occupying and entertaining bored or stressed minds. (1) Museums are not places for communicating a lot of factual information to a lay public, but they do present opportunities for awakening people's interest in a subject, so affecting their educational desires. (2) Museums, because they are *convivial* institutions in which things that are not immediately understood can be enjoyable challenges, could motivate whatever learning visitors wish to engage in.

The first conclusion is unexceptional and undoubtedly enjoys wide support. But it is not free of problems because no one has shown so far what it means in terms of practical exhibition design. However, Grunig's theory and empirical results from research in museums indicate at least some of the factors that have to be taken into account. These are equally important in connection with the second conclusion. Thus, if museums are going to design exhibits that facilitate information processing, they must above all begin to pay attention to their visitors' concerns, interests, expectations and so on, and make an effort to speak to them in a familiar language. This is what we have been trying to do in recent years at the Natural History Museum (e.g. Miles 1986b).

Acknowledgements

I am indebted to my colleagues Dr Giles Clarke, Ms Joanna Jarrett, Ms Deirdre Janson-Smith and Mrs Anita Morris for their comments on my first draft.

References

Alt MB 1980 Four years of visitor surveys at the British Museum (Natural History) 1976–79. Museums Journal 80:10–19

Alt MB, Shaw KM 1984 Characteristics of ideal museum exhibits. British Journal of Psychology 75:25–36

Bodmer WF, Artus RE, Attenborough D et al 1985 The public understanding of science. Royal Society, London

Borun M 1977 Measuring the immeasurable: a pilot study of museum effectiveness. Association of Science-Technology Centers, Washington DC

Danilov VJ 1982 Science and technology centers. MIT Press, Cambridge, MA

Falk JH 1983 The use of time as a measure of visitor behaviour and exhibit effectiveness. Roundtable Reports: Journal of Museum Education 7:10–13

Falk JH, Koran JJ, Dierking LD, Dreblow L 1985 Predicting visitor behaviour. Curator 28:249–257

Griggs SA, Alt MB 1982 Visitors to the British Museum (Natural History) in 1980 and 1981. Museums Journal 82:149–155

Griggs SA, Hays-Jackson R 1983 Visitors' perceptions of cultural institutions. Museums Journal 83:121–125

Grunig JE 1979 Time budgets, level of involvement and use of the mass media. Journalism Quarterly 56:248–261

Grunig JE 1980 Communication of science information to nonscientists. In: Dervin B, Voigt MJ (eds) Progress in communication sciences. Ablex Publishing, Norwood, NJ, p 167–215

Grunig JE, Disbrow JA 1977 Developing a probabilistic model for communications decision making. Communications Research 4:145–168

Hudson K, Nicholls A 1985 The directory of museums and living displays, 3rd ed. Macmillan, London

Kemp G 1986 On with the show business. The Sunday Times (London) (5 Jan)

Kimche L 1980 American museums: the vital statistics. Museum News 59:52–57

La Follette MG 1982 Science on television: influences and strategies. Daedalus 1982:183–197

Laetsch WM 1979 Conservation and communication: a tale of two cultures. Southeastern Museums Conference Journal p.1–8

Lewis BN 1980 The museum as an educational facility. Museums Journal 80:151–155

Linn MC 1976 Exhibit evaluation: informed decision making. Curator 19:291–302

Loomis R 1973 Please! not another visitor survey. Museum News 52:20–26

Melton AW 1935 Problems of installation in museums of art. Publications of the American Association of Museums NS14:1–269

Miles RS 1986a Museum audiences. International Journal of Museum Management and Curatorship 5:73–80

Miles RS 1986b Lessons in 'human biology': testing a theory of exhibition design. International Journal of Museum Management and Curatorship 5:227–240

Screven CG 1974 The measurement and facilitation of learning in the museum environment: an experimental analysis. Smithsonian Institution, Washington DC

Shettel HH, Butcher M, Cotton TS, Northrup J, Slough DC 1968 Strategies for determining exhibit effectiveness. American Institutes for Research, Pittsburgh

DISCUSSION

Dixon: You have put on more didactic exhibits in the Natural History Museum recently, such as those in the Hall of Human Biology. At the same time you say there is no point in trying to assess how much people have learnt when they leave the museum. I am puzzled by that.

Miles: The Hall of Human Biology is an attempt to do something about the problem of communicating through exhibits, as is most of our work over the last nine or ten years. My generalizations refer to large, traditional science museums as a group.

Friedman: The place to look for the data is in Chandler Screven's bibliography (1984) and the papers listed there. My personal summary is that a small but crucial amount of learning takes place, with visitors getting excited about perhaps one thing out of 150 presented. That small piece of knowledge can, however, produce a large affective change in the visitor. Museums can awaken a desire for education through this small piece of learning or interest.

Thirty days after the New York Hall of Science opened we had 10% repeat visitorship, which suggests that people are having a rewarding experience. We have no evidence that the effect of obtaining a crucial bit of knowledge extends beyond encouraging a repeat visit but I want to look for additional effects.

Falk: No visitor reads all the labels but every visitor reads some labels and finds something of interest. But if you try testing visitors as they leave, nearly everyone fails. On the other hand if you asked individuals what they gained in knowledge and allowed them to answer in their own individual way, based on their own expectations and background, you would find there was a gain. The tests follow the bias of more formal education. Tests assume that exhibits are part of a curriculum and they try to find how much of the curriculum was learnt. That doesn't work.

Dixon: There must be people who come out of a museum after seeing for the first time the difference between, say, the concept of a molecule and a gene. Is there no way you can tell us about that event?

Falk: No. You have to find the subset of the museum-going population who have had that experience.

Laetsch: A common question museum visitors ask is 'Is it real?'. Many people go to museums and zoos to see an object or animal that is real, not to learn something new. Seeing a real zebra or a camel is important for children in defining their world.

Tisdell: The opportunity to see the Rosetta Stone in the British Museum improves the credibility of Egyptology. Seeing the real thing is indeed important.

Screven: Academic-type tests have been used to measure learning outcomes of science exhibits because these define what museum scientists say reflect the messages their exhibits are trying to communicate—which is not surprising since museum scientists are academically oriented and view museums somewhat like three-dimensional libraries.

But the museum exhibit potentially is a powerful teaching tool. Its focus on objects and its three-dimensional, life-like settings attract visitor interest and potentially can generate interest in learning more. But, learning more what? It probably is true that many objects and phenomena displayed in science exhibits not only attract attention but can motivate viewers to seek out text and diagrams that promise explanations of what they are seeing or answers to questions about these objects. If so, this look-then-read relationship has promise because it sets the occasion for learning more about things. It is very important that when visitors read labels they find meaningful information that directly relates to the exhibit experience or to the questions generated by it. Otherwise, visitors simply stop reading! If we are serious about teaching science in public exhibit environments, we need to know more about when, where and how to provide explanatory supporting information that will be effectively used by visitors. Existing research on labels is not very helpful, though this research picture should improve in the next few years.

Evered: Roger Miles said there was a bimodal distribution in his figure (p 119) showing how long visitors spent in front of exhibits (Fig. 1), but the figure only showed that the distribution was heavily skewed. It was not possible from the way the data were presented to know whether the distribution was bimodal since the observations on the right side of the diagram were aggregated. Was the distribution in fact bimodal or merely skewed?

Falk: That figure is based on my data. It is bimodal in the sense that the data represent two populations, but to emphasize my point I lumped the extremes. Whether one spends two minutes in front of an exhibit or three minutes is irrelevant when these times are compared to the mean of 30 seconds per exhibit. Means are usually based on normal distributions and the mean represents the mode. But in the museum these distributions typically represent about 10% of the visitors. The distribution is bimodal in the sense that most visitors spend either very little time or an inordinately long time in front of an exhibit.

Serrell: I would encourage Roger Miles to keep going in the direction he described. Evaluation efforts might simply concentrate on two questions: 'What was this exhibit about?' and 'Can you give me one example of a new idea you are walking away with?' If visitors can tell you in a general way what the exhibit was about you have got your main message across. The answers to the second question will give you something specific and factual.

Evered: If you also disturb people's preconceived notions of the ways in which the natural world works you may not be able to measure that effect but it probably has some long-term educational benefit for the individual concerned.

Tisdell: You might ask visitors whether they would be interested in reading more on these things in the future, to test their curiosity. Some people might indicate that they would have an interest but later you might find that they don't follow it up.

Lucas: You can get some of that evidence from what people buy in the bookshop before they leave.

Laetsch: We are telegraphing our academic bias here by insisting that there has to be an intellectual gain or reason for going to a museum. If we destroyed all the museums in the world tomorrow, people would demand that they be re-created in a similar form. In the United States people from across the socioeconomic scale vote for bond issues and other schemes for building museums and science centres and zoos. In other words they demand that those institutions should be there for the purposes they serve now. Museums and similar institutions are not regarded as academic institutions but they are regarded as being very important.

Miles: The British government is running a large-scale test on that point, proportionately reducing its contribution to museum funds and encouraging entrance fees to be charged. The public will vote 'for' by paying, and 'against' by staying at home.

Miller: It is legitimate to expect some behavioural change after a visit to a museum. Otherwise we could franchise them to Disney. If we want people to carry away something, we ought to be able to define what that is and measure it. It is reasonable to expect such visits to stimulate other kinds of information acquisition. It should be possible to telephone a random sample of visitors and find out what they are doing three months or six months later that may have been influenced by their visits to museums.

Falk: There is a value in reaffirming and extending the concepts that most of us have been forced to deal with in a very flat two-dimensional and rhetorical way. We have been cut off from a lot of the reality of life. These institutions permit a re-establishment of what we have been told is reality but have not seen for ourselves. Museums present the concrete abundantly well but they present the abstract abysmally. If you have to depend on a label to explain something abstract, such as the theory of relativity, you are in trouble, since most visitors cannot be depended upon to read the label. Hence this raises a paradox for science museums in particular, which are attempting to describe a topic—science—much of which is extremely abstract.

Screven: What Jon Miller said is important. There is no hard evidence, but it seems likely that museum visits may have important effects after visitors leave the museum that are reflected in school, family, leisure activities, vocational choices, and so on. For example, a good experience with a well-designed archaeology exhibit may result in a visitor later attending to a magazine article on archaeology while waiting in a doctor's office, or to choosing archaeology as a topic for a term paper.

Laetsch: We shouldn't sweep Disney under the rug. Every country that plans new science museums or science centres sends groups to visit Disneyland and Disney World in Florida, particularly the Epcot Center there, which is semi-educational in intention.

Miller: I don't know anyone who has found the Epcot Center scientifically enlightening.

Tan: We send our officers to Epcot not for scientific education but to study techniques for attracting visitors' attention. Any organization, commercial or otherwise, that wants to sell something has to get the potential buyer's attention first. That is why I like the idea of an exploratory. Once you get somebody's attention or interest, you can proceed to sell the scientific messages, including the more abstract ideas.

Duensing: We have found that television is a very powerful medium in attracting attention or interest in the Exploratorium. Several years ago a programme called 'The Palace of Delights', about the Exploratorium was made for the *Nova* science television series. Whenever it is shown, our attendance rises dramatically, perhaps more than through any other marketing tool we use. It is interesting that the passive entertainment medium of television can be used to elicit a desire to visit a very active museum.

Caravita: Outside schools we maintain that entertainment, interaction with concrete objects, and a wide display of objects are important. But for in-school learning we shift to different theoretical models for learning. Out-of-school education is run by scientists rather than by educationalists. Museums and other such institutes can perhaps provide schools with different educational strategies to replace their more formal and traditional methods.

Tan: The traditional view was that if it is fun it cannot be educational, but learning through funs seems to be the key phrase now. The key to learning is not how much we can tell the public but how to get schoolchildren to understand that learning can be fun. The teacher who tells them they are not visiting our Science Centre to enjoy themselves but strictly to learn is the weak link who kills the curiosity every child has.

Laetsch: The most critical variable is certification—the need to have a piece of paper ostensibly proving what you have learnt. That is the barrier that almost everything bangs up against.

Miles: The point about schoolteachers applies to museums as well. If you are providing information for people who want it for instrumental purposes, you don't have to go to much trouble in the presentation. A classic example is the scientific paper. However, if you are trying to attract attention in an informal environment the affective aspects of the design become very important. Traditionally museums, at least in Europe, were designed by curators with a scholarly background who assumed that information was going to be eagerly sought by visitors on their (the curators') terms. This was a false assumption. Nowadays all museums that seriously care about making contacts with their visitors make great efforts to attract them, interest them, and hold their attention.

We are still left with the major problem of deciding just what is achievable in science communication in museums and under what sort of conditions. That could be considered independently of the problem of how we evaluate what is going on.

Quinn: We have a tendency to stick with a lot of our traditional techniques of presenting information. I strongly support looking outside to see what effective techniques are being used in the entertainment industry and learning to adapt them to the museum setting. Education and entertainment can go hand in hand. Some of the most effective teachers in the world are some of the best actors and entertainers, and conversely. Museums need a spectrum of techniques for presenting information to visitors.

I am not sure that we can test what museums do for people by phoning them six months later. Some things are easy to test but many are not. I wouldn't want museums to do only what is easy to test just so that we can say we are being effective.

Macdonald-Ross: It is self-evident that researching existential outcomes is difficult. Researching the consequences of reading novels is similar. It is possible to do it but most research groups cannot find the funds or the staff.

Tisdell: Some visitors said there was not enough information about the exhibits in the Natural History Museum, others said there was too much. People have different degrees of interest so the information ought to be presented to cater for both groups.

Miles: For various reasons we cannot exactly match the variety of our exhibits to the variety of interests of the visiting public. However, we do design some exhibitions for people wanting more detailed information, as we do individual exhibits in more popular exhibitions; and we are experimenting with ways of presenting information at different levels throughout exhibitions. There is always more to say than is possible in an exhibition, and the selection of content is best guided by a clear statement of purpose. In this connection we go to some trouble to define our target audiences well.

Laetsch: In the academic world, if something is popular it is not regarded as rigorous. On my campus, if someone writes a book that turns out to be popular other faculty members call that person a journalist. That attitude goes all the way through the system.

Miller: Popularity does not necessarily mean that something has to be poor science. But popularity by itself does not mean effective science teaching. The methods in the Epcot Center may be quite good but I don't think the Center teaches science very well.

Miles: There are two sorts of organizations: those that are in the business of profit and fun and that claim to be educational to give themselves some respectability, and museums, which are in the business of education but need to provide fun to attract their audiences and increasingly nowadays need to earn money. We have to retain a clear sense of our priorities if we are not to slip into Disneyland.

Laetsch: We are trying to discover how to get through to the non-attentive group. If we do get through to them we have something that is popular.

Screven: Thomas Malone reported (1981) that the three characteristics of the most popular video games were fantasy, challenge and curiosity. These motivations might be harnessed for instruction in science.

Miles: We have a genetics programme running in the Natural History Museum at the moment which was guided by Thomas Malone's work and seems to produce good results.

Gregory: In a lecture or a television or radio programme one has to structure the facts in order to make points about the history of technology, or whatever the subject is. In the museum, in some sense you have the raw data and you can look at it in lots of different ways. If you know about the history of technology a lot of things go through your mind when you look at something like the Watt beam engine. Captions don't help somebody who already knows about the subject, and they might deflect somebody who does not know about it into a trivial or boring way of thinking. This is an immense problem. In a television or radio programme you can present one of the facets but in a museum you can't. The museum needs to be neutral in order to be fair as a data source. Yet as soon as you structure it you distort it, in some ways, for the visitors.

Caravita: You might offer different types of structures rather than neutral structuring for the same objects. This might help to show how science provides different frameworks to interpret the same phenomena.

Gregory: I agree, if it is practical to provide alternative structures. Alternative structuring is done in some cases, for example in work sheets provided for school groups. In the Science Museum in London the objects are arranged according to their technological function. Work sheets direct people to different galleries to look at the social implications of those objects in the Industrial Revolution. The real problem is that, if I don't know I want a particular view of the object, how am I given an entry into it? How can I be shown that I can look at these things in this sort of way?

Friedman: In the New York Hall of Science we are going to experiment with an electronic label that may answer that question. Visitors will be able to choose different points of view from a menu.

Duensing: We divide our exhibit labels into three sections. First we have a section called 'To do and notice', about how to get the exhibit to operate. Then comes a description of what is going on, which might be a technical explanation in what we hope are simple, jargon-free terms. Then in our more recent labels we add a third component called 'So what?', which could be a picture or more words about how this exhibit relates to the outside world. We don't ask questions very often. Instead we say things like 'Notice what happens when the light overlaps', which is less threatening than a question. Visitors don't feel there is a 'right' answer to find; their observations are what is important.

Screven: The use of questions in museum exhibits has attracted a lot of interest. Using questions to facilitate learning is not new, but exhibitors are only now learning how to use questions to improve usage of exhibits by viewers. Questions encourage participation. They can be parts of headings, labels, embedded in guidebooks, or serve as components in interactive devices such as flip panels, punchboards, and computers. Given poor background on many exhibit topics, visitors do not process exhibit information effectively: they have difficulty distinguishing between the trivial and non-trivial, the important and unimportant. Questions can focus attention on key distinctions, similarities, alternatives, misconceptions, and attitudes, and encourage visitors to discover for themselves conclusions, generalizations, and principles (Screven 1986). Questions can direct viewers where to look or what to look for, encourage comparisons, and encourage them to look for processes and similarities. If a relationship between A and B is important, questions can help viewers 'discover' the relationship in the exhibit.

It is fairly easy to get people from 10 to 70 years of age to cooperate—even without gadgetry; for example, the flip question is one of several approaches to non-automated, interactive question formats (Screven 1986). Viewers choose one of several hinged panels depicting alternative photos, objects, or statements and they receive answers, comments, or other feedback when they lift one of the panels. If questions are specific and it is clear what to do or where to look for information, visitors (children and adults) normally do not guess.

Such non-automated methods are useful for building simple concepts by helping to interface visitor attention with exhibit content. They probably are *not* as productive for more complex concepts and when there are large individual differences in visitor interests, background, and needs on a given topic. In this case, computers with branching capabilities are probably better because they can adapt to visitor interests and backgrounds. Computers are also useful when greater visitor effort over time is required. The electronic revolution has immense potential for education everywhere (Lepper 1985). If museums have a potential as alternative environments for the communication of science, I think the potential for computers in museums is just as great. Computers provide a practical way for exhibit staff to adjust to visitor needs and to changes in the readiness of visitors for types and levels of information as their understanding progresses through subsequent visits. Some applications of computers in museums have been described by Friedman (1983) and myself (1986), among others.

Friedman: Schoolteachers complain that they cannot compete with the entertainment provided by television programmes and museums. 'I can't sing and dance and change colour' is a typical comment. In museums we want to make visitors feel confident in their own ability to learn but sometimes we are doing the opposite: teachers feel they cannot produce the same effects.

Rhodes: If a teacher with all the advantages of personal communication cannot beat a box in the corner which cannot interact with its audience in any way, he or she has a real problem. Of course teachers can compete if they try.

Caravita: They have to compete not on the basis of their authority but on the basis of their professional ability.

Lucas: The purposes of teaching are different from those of the media. The teacher is trying to build a deeper understanding and not just present the surface of the phenomena. The museums and the media put the emphasis on the phenomena.

Rhodes: We are all saying that museums, radio and television are not good on facts. They are good only at exciting people's interest. Teachers have to take advantage of that interest once it has been aroused.

Laetsch: Museum and zoo visitors have about 50 minutes to interact with something. What does that say about the goals that many of these institutions express? We can reach large numbers of people about matters related to science and we can even interest them in these. Later we should try to come to grips with what this means in relation to the expressed goals of scientific literacy.

References

Friedman A 1983 The new technologies and museum education. Roundtable Reports 8(5):12–14

Lepper MR 1985 Microcomputers in education: motivational and social issues. American Psychologist 41(1):1–18

Malone TW 1981 Toward a theory of intrinsically motivating instruction. Cognitive Science 4:333–369

Screven CG 1984 Educational evaluation and research in museums and public exhibits: a bibliography. Curator 27(2):147–165

Screven CG 1986 Exhibitions and information centers: principles and approaches. Curator 29(2):109–137

Science centres and exploratories: a look at active participation

SALLY DUENSING

The Exploratorium, 3601 Lyon Street, San Francisco, California 94123, USA

Abstract. There is an increasing trend in science museums world-wide to present aspects of science and the phenomena of nature in ways that encourage individual investigation and discovery. Museum visitors from a broad range of backgrounds and interests become active participants in these centres. Together the overall organization and the individual exhibits of the museum create a stimulating, involving environment. At the Exploratorium in San Francisco the exhibits are seen as interconnected chains of curricula. There is no correct place to begin or end in the different exhibit areas. Visitors are encouraged to wander. Teachers select pathways through the exhibits for their students based on what they are studying. Concepts are presented in multiple contexts. This enables the exhibits to be used as effective props for many different levels of teaching and learning. Visitors can build up a good framework for understanding a concept through experiencing it from a variety of viewpoints.

Exhibits themselves provide a wonderful vehicle for encouraging people to ask 'I wonder what would happen if. . .'. They can then try to find out. There are a variety of things to see or do at any one exhibit. Through active participation, science can be portrayed as a lively process rather than a static collection of facts. In addition, encouraging people to fiddle about with nature can arouse curiosity and instil the pleasure of learning, of finding something out for yourself. And this can create a motivation for further learning and a feeling that the world is understandable.

1987 Communicating science to the public. Wiley, Chichester (Ciba Foundation Conference) p 131–146

For the past 20 years or so there has been an increasing trend in science centres and museums world-wide to present aspects of science and the phenomena of nature in ways that encourage individual exploration and discovery. This process of exhibitory is often called participatory, interactive, or hands-on (Gregory 1983, Tressel 1984). The rationale or interpretation of what this process is all about is quite diverse in the museum community, and at times is oversimplified. In this paper I use the history and experience of the Exploratorium, San Francisco, to first define and then explore some of the significant features and goals involved in creating a museum that encourages participation.

Active participation is a process of teaching but is not the subject matter itself. This notion sometimes gets lost as people talk about creating interactive exhibits and museums. People tell me they are building a museum 'like the Exploratorium' but when I ask what is going to be exhibited, they reply 'Oh you know, that hands-on stuff like you have'. Active participation is an outcome of a series of pedagogical concerns and goals. It is not the fundamental principle in developing the content of the Exploratorium; rather it is the behavioural outcome. Our starting point is to provide visitors with access to direct experience of the ideas and phenomena of science.

In 1969, Frank Oppenheimer began the Exploratorium with the goal of creating a museum that would encourage people to fiddle about and notice natural phenomena as a way to learn science. During a visit in 1965 to the London Science Museum, Frank realized that the United States was lacking in museums of basic science. There were museums of technology but very few institutions that presented basic science.

'On the whole', he wrote (Oppenheimer 1968), 'people have very little opportunity to have any direct experience with the separate elements of nature or technology. They watch ocean waves, but have never been shown how to observe the way waves pass through each other, bend around corners or bounce off cliffs. In a science museum, one can provide these direct experiences with the behaviour of light, sound and motion. One can set up these experiences in such a way that they not only generate, but also partially satisfy curiosity. Science is not just a process of discovering and recording natural phenomena; it is a process which develops our ways of thinking about nature and which enables us to find the connections that simplify and at times enrich our comprehension and awareness of nature'.

Direct experience was and continues to be the basic tenet that the Exploratorium follows in trying to present ideas of science. It is through the public's direct experience with natural phenomena that active participation occurs in the museum (Fig. 1). We have tried whenever possible to exhibit a natural phenomenon rather than an explanatory model. Often a model is an abstraction that is only exciting to someone already familiar with the idea or concept. There is something intrinsically interesting in the 'real thing', be it the shimmer of a rainbow on a wall, the sound of a real musical instrument, or a telescopic view of real stars. Affording the public the opportunity to experience the real thing is perhaps one of the most significant features that a museum can provide. This adds to its ability to interest a broad range of people. Even physicists who visit the Exploratorium are delighted by seeing things they already know but are experiencing in a new or unusual way. I have often heard the comment, 'You know, I've read about this but I've never seen it before.'

We often get adult visitors who say, 'Gee, I wish science had been taught like this when I was a kid.' What they probably mean is that after rejecting

FIG.1. *Tornado*. Water mist is formed into a vortex through cross-currents and an overhead fan that creates an updraft. This young visitor can touch the tornado and can also block the wind to see the shape disappear.

science most of their life they now feel that they could, in fact, have understood it and been interested in it if it had been presented in this way. As Frank Oppenheimer once said, 'Conveying to our visitors a sense that they can understand the things that are going on around them may be one of the more important things we do.'

Through the theme of perception we have been able to show some of the interconnections between disciplines of science; for example, finding the connections between sound and hearing can then lead into an investigation of waves, and then, perhaps, to certain aspects of electricity. Use of perception as a starting point encourages visitors to participate actively in many of the concepts demonstrated in the museum.

The theme of perception has been very effective in getting visitors involved in investigating the process of science rather than looking for the 'right answer' (Fig. 2). We are more concerned that our visitors gain confidence in their own observations than that they should walk away with some fact or principle.

Overall museum organization and active participation

For the term 'active participation' to have any real significance in a museum, it has to refer to both the overall organization and the design of the individual exhibits. There are two basic precepts in the overall organization of the Exploratorium which I feel have been quite influential in creating a stimulating, involving museum environment. One is the use of perception as the overall guiding theme in the Exploratorium. The other is that we have never tried to be an encyclopaedia of science.

Too often, science museums try to cover too much, using too few examples. For instance, one or two examples of reflection, refraction and colour mixing are presented and it is thought that the behaviour of light has been covered. While these individual exhibits may be quite interesting and interactive, they can provide no more than an introduction to the behaviour of light. This superficial coverage limits the amount of participation and learning possible for the museum's visitors.

In providing a rich environment of direct experiences for our visitors we have limited the areas of science covered in order to present an array of examples on any one concept or idea. In our section on images and refraction, for example, a visitor might start by looking at how light is bent by lenses, and then discover that water of different temperatures can bend light in much the same way. Or the discovery might be made the other way around—beginning with water and ending up with lenses. A visitor who dips Pyrex and flint glass rods into immersion oil to see that the Pyrex glass rods seem to disappear might end up investigating other aspects of refraction, or become interested in his or her perception of boundaries.

FIG. 2. *Gray Step I.* Direct experience of how important edges are in perception occurs in the *Gray Step I* exhibit. When the rope covers the boundary between the squares (Fig. 2a), no difference is perceived between them. Move the rope and a sharp contrast is visible (Fig. 2b). Note that both squares are actually identical, but not a uniform shade of grey. The left sides of each square are the brightest part and each square gradually grows darker towards the right. (© The Exploratorium; Susan Schwartzenberg.)

The exhibits at the Exploratorium are used by families in a very casual 'Sunday sightseeing' sort of way, but they are also used by college professors as teaching props for their physics instruction. Elementary and high-school teachers select pathways through the exhibit areas for their students, based on what they are studying. This type of varied use would not be possible through the selective presentation of single examples of any one idea.

We see our exhibits as interconnected chains of curricula that can be and are used for many different purposes. The arrangement of the exhibit areas in the museum also contributes to the encouragement of individual participation. At the Exploratorium, there is no one correct place to begin or end. Visitors are encouraged to wander, to find things on their own. People can go at their own pace, stay as long as they want investigating any one idea. In 1973 Albert Parr, from the American Museum of Natural History, saw this need and wrote: 'What we need more than ever before is some wilderness areas (in museums) for all ages, where rich and enjoyable stimulus fields invite our bodies and our minds to explore at will without other direction for every step we take and every thought we have on the way. We must offer a wide choice of paths through the landscapes of our exhibition halls to challenge our visitors' curiosity and their personal sense of sequence.'

In providing multiple examples in a variety of contexts of interesting or important phenomena, the Exploratorium recognizes the individual differences of learning in our visitors. If one of the exhibits does not make sense or seem very interesting, a nearby one may offer more interest and lend more clarity to an idea. As far as we can see there is no age, training or cultural limit to the range of people who enjoy the Exploratorium. Pre-school groups, senior citizens, disabled people and gifted people use the place with equal satisfaction. It is ridiculous to think that every exhibit should appeal to every person. Providing multiple examples enables the museum to be a setting for teaching and learning on a wide variety of levels.

Exhibits and active participation

A number of factors contribute to making individual exhibits participatory in nature. Perhaps the greatest over-simplification in defining what 'active participation' in an exhibit means is the notion that it means touching; that the act of touching something in and of itself creates a stimulating, involving exhibit. Many of the Exploratorium's exhibits, especially the visual perception exhibits and sound exhibits, have nothing to do with touch. In a sense they are hands-off, but they are quite involving and encourage a great deal of investigation and participation. Generally, active participation means the process of allowing the visitor to change and explore some of the characteristics of the phenomena being exhibited. Visitors become involved in the visual perception exhibits at the Exploratorium through using themselves as the agents for

FIG.3. *Light Island*. With this exhibit visitors can explore many different aspects of light. Using lenses, mirrors and coloured filters the light beams can be bounced, bent and mixed to create beautiful patterns.

change. They look at something differently, close one eye, change their point of view, compare their perception with that of friends. They explore something on their own, raise their own questions and often come to their own conclusions about the phenomenon in the exhibit.

The starting point for all of our exhibits, as they are created, is our consideration of what we want to teach and how we can teach it in an interesting way. Exhibits are most often developed by people who are interested in the phenomena to be displayed. In creating an exhibit we explore what is fun to do and what is intriguing about the phenomenon. We try to build exhibits that please our visitors, as well as ourselves.

Most exhibits are very flexible in how they can be used, and they often include a variety of things that can be changed to affect the phenomenon being demonstrated. Our Light Island exhibit, for example (Fig. 3), has three or four flat mirrors, four different coloured filters, and one curved mirror lying loose on the table; a prism, a convex lens and a concave lens are also part of the exhibit, attached by cables. Visitors can use any one of these lenses and mirrors individually or in combination to see how they affect the light beams on the table. This exhibit has proved quite interesting to our visitors because there is so much for them to do. They can create a myriad of patterns and colours of light as well as do a systematic study of optics. I have seen kids as young as seven or eight spend 15 or 20 minutes exploring patterns, reflections and images.

Another example of flexibility is in the exhibit called Vocal Vowels (Fig. 4). The main point of this exhibit is to show how different-shaped chambers of the throat and mouth affect the sounds of speech. A multiple-frequency sound source (we use a reed from a duck call) is placed at the opening of each chamber. A visitor can pump a bellows and hear the buzzing sound of the reed turn into the sound of a vowel as certain frequencies are amplified or cancelled by the chambers. This is an interesting phenomenon in and of itself, but we wanted to add another component that would allow the user to experiment with the shapes of the chamber to see what happens to the sound. We added a chamber with movable slats so that a wide variety of shapes could be created. The main idea of this exhibit is speech physiology and is covered quite well through the fixed chambers; adding the flexible chamber allows for much further experimentation. Exhibits like this provide a wonderful vehicle for encouraging people to ask, 'I wonder what would happen if . . .' They can then try to find out.

We do not insist that people only get the main point of our exhibits. Many of the activities we suggest that visitors should try are relevant to the main idea, but the exhibits themselves allow for a great deal of fiddling about. At the exhibit Two As One, we create a three-dimensional effect by presenting each eye with slightly different pictures, thereby showing that binocular disparity can create the perception of depth. The exhibit has many different

FIG.4. *Vocal Vowels*. When a reed is placed in front of the clear chamber the buzz sound of the reed gets transformed into the sounds of different vowels. Each of the chambers mimics the shape of the throat and mouth when different vowel sounds are spoken. (Photograph by Esther Kutnick.)

cards. For example, visitors can look at cards with lines and cards with circles in different positions that create the illusion of depth. We also have cards that present two very different images to each eye so that visitors can experience the effects of eye rivalry—two different faces can be combined, letters can be viewed so that whole words appear when viewed binocularly. All of the cards can be mixed and matched, enabling the visitor to try any number of possibilities.

Push-button exhibits are often referred to as 'hands-on' or 'participatory' because they involve the visitor somehow activating something with their hands. The problem is that most push-buttons leave too little for the observer to control. While the display is going on there is not enough connection between the apparatus and the phenomenon. As a result, the idea presented is often unclear. And finally, once you have seen the effect, there is nowhere else you can go with the idea.

Providing many choices at an exhibit makes it possible for a visitor to figure out his or her own experiments. The freedom to experiment also includes the freedom to fail, to make the exhibits misbehave. Things can be taken past their functional limits so that the range of how or why something works can be experienced.

In almost all forms of learning it is as important to appreciate what does not work as to appreciate what does work. Even a simple phenomenon, such as the formation of an image by a lens, is difficult to appreciate unless one can twist and turn the lens, move it all about, or put various things between the lens and the image. Such detailed control does not rule out remote control, although whenever possible it is much more interesting to allow visitors themselves to change what is happening or stop something from working.

Several months ago I saw someone at our Catenary Arch exhibit. This exhibit has 21 loose blocks that can be arranged into a catenary arch that will stand on its own when built correctly. I noticed that the way the visitor was arranging the blocks would not work. I went up to him and started to tell him the 'right' way to do it. He turned and said to me, 'I know how to do it right, I want to do it wrong to see what will happen.' He was, of course, quite right. He reminded me how important it is to allow people to find out things for themselves.

Several years ago we held a conference at the Exploratorium to gather information and suggestions that might help us to build successful electricity exhibits. Participants included museum professionals, exhibit designers, scientists and educators from many different institutions. As a way to think about the important features our exhibits should contain, we asked participants to recall how they became interested in electricity. Everyone had an amazingly similar story. Their involvement grew through play: play with toy trains, watching the transformers heat up, making sparks as tinsel fell on the tracks or taking various devices apart to see how they worked. Playing was cited by all the participants as crucial for their later understanding of the concepts presented in school. People said they had some context for the ideas, some reason to care. Most said that if someone had interfered with their play to explain the phenomenon they would have been turned off immediately. They said it was important to have the experience to see what would happen, to just fiddle with the things, to let the questions emerge on their own.

Another important aspect of building participatory exhibits is the social consideration. Museums are social places and exhibits can facilitate social interaction to a remarkable degree. In the Report on the Evaluation of the Test Bed, conducted at the Launch Pad exhibit in the London Science Centre (Tulley 1985), the exhibit selected as the most interesting by many of the visitors interviewed was called the Grain Pit. The visitors who chose this exhibit as their favourite all commented that there was a lot to do and that it involved people working together.

We have noticed at the Exploratorium that not only is it fun for people to do things together at an exhibit, it is also fun to watch others. We try to build our exhibits to be free-standing so that more than one person can gather around the exhibit to experience the phenomenon themselves or to watch how other people use the exhibit. This also encourages others to get involved: it is quite common to see people gather around an exhibit to watch what another person is doing. You often hear people yell to a friend or family member, 'Hey, come here and look at this!' An amazing amount of teaching and learning occurs where someone shows another person what they just saw or experienced.

Exhibits can elicit a new interest in science. A dramatic example of this happened when an Australian physics professor, Dr Michael Gore, visited San Francisco in 1975 and brought his family to see the Exploratorium. He had to, as he said, drag his family in and then, three hours later, drag them out. Much to his surprise, his family was as interested and involved in the exhibits as he was. This was no small matter. For years he had tried unsuccessfully to interest his family in some of the very same ideas that were shown in the exhibits that they had played with at the Exploratorium.

Dr Gore was so impressed by this form of teaching that on his return to Canberra he started his own science centre called Questacon. Questacon was recently designated as the model for the new science museum of Australia.

Conclusion

In this paper I wanted to convey the importance of the whole museum as well as the individual exhibits in creating an environment that encourages active participation in a thoughtful and meaningful way.

Museums have the unique ability to provide direct experience with real things, presenting possibilities that are difficult if not impossible to achieve in other forms of teaching. Museums can present broad overviews as well as detailed studies of concepts and ideas. What we communicate are ways of thinking about nature. The richness of science can be communicated by multiple examples which allow the visitors to explore an idea from a number of different viewpoints. Too often, people think that science has to be diluted in order to be interesting to a broad audience. At the Exploratorium we have found the opposite to be true. This richness enables the museum to be interesting and accessible to multiple age groups and levels of interest.

Although all museums are based on props, science centres and exploratories are basically museums of ideas. Rather than collections of artifacts, they are collections of phenomena. The trend of creating more active museums is a way to make available more direct experience with the ideas, phenomena or objects displayed. Our aim is to instil in our visitors a feeling of connection and caring, not just a feeling of awe.

Through active participation, science can be portrayed as a lively process rather than a static collection of facts. Encouraging people to fiddle about with nature can arouse curiosity and awaken the pleasure of learning and the satisfaction of finding out something for themselves. This can create a motivation for further learning and a feeling that the world is understandable.

References

Gregory RL 1983 The Bristol Exploratory—a feeling for science. New Scientist p 484–489 (17 November)
Oppenheimer F 1968 The role of science in museums. In: Larrabee E (ed) Museums and Education. Smithsonian Institution Press, Washington, DC, p 167–178
Parr AE 1973 Refuge from other-direction. Discovery (New Haven) 9 (1):34–36
Tressel GW 1984 A museum is to touch. In: Calhoun D (ed) 1984 Yearbook of science and the future. Encyclopedia Britannica, Chicago, p 214–231
Tulley A 1985 Report on the evaluation of Test Bed 2. London Science Museum, London

DISCUSSION

Gregory: The Exploratory in Bristol owes a lot to Dr Oppenheimer too. I was immensely impressed by his work, and in Bristol we are trying to do the same things as you do in San Francisco.

In a certain sense 17th century science threw the observer out in its search for objectivity. The last thing that was wanted was human fallibility. Now we are trying to bring the observer back into science and show what it is to observe, make decisions and understand. We are helped by the philosophy of artificial intelligence and its discussion of how one makes a machine that can see, make decisions, and understand the world around it. Some philosophers argue that no machine can ever understand, whereas a brain can. This is rubbish. Information processing is the key and one can show that there is a series of processes. In the Exploratory we start with the observer and end up with robotics.

In Bristol we want to encourage international competition in robotics, in games such as snooker or billiards. We want to make games respectable in science.

One example of what we do in the Exploratory is the use of pendulums to show the limits of what can be done by the eye and the ear in perception. We only bring in electronics when it is necessary. This shows the relationship between the human observer and instrumentation. The interplay of human judgement and instruments is a theme that relates theoretical science to technology and human judgement quite neatly. Thus, you can imagine you are space travelling with a pendulum, but it is more than a game and it isolates and reveals principles in a way that could not be done in a book. As you play with phenomena and so on, you learn about yourself because you are making judgements and some fail and are surprising.

We have a lot of colour-mixing demonstrations that relate to technology and to the human eye. They can tell you how television works, how printing works, and so on.

Lucas: How important is that pervading theme of perception in the Exploratorium? If such a theme is important to provide the unity, what about other aspects of science?

Duensing: Perception is a very important theme for the Exploratorium. It not only links many different areas of science, it has also enabled us to integrate aspects of the arts and humanities with science in very meaningful ways. For example, music and language exhibits are integrated with exhibits on sound and hearing. The way an artist looks at nature can be quite different from the way a scientist looks at it. As both artists and scientists are involved in exhibit development at the Exploratorium, the experience of both is presented, which provides a much broader view of nature. While perception is a very good vehicle for this, I am sure that it is not the only theme that can accomplish this. Part of the importance of having a theme is that it provides a focus for the museum, and this makes a museum more effective in communicating ideas or concepts. Exhibits can work together to help the visitors see patterns in seemingly unconnected phenomena or events.

Lucas: You mentioned that people don't see the labels for those areas.

Duensing: The labels I was referring to were large signs that are suspended over the different exhibit areas in the museum. Most people do not notice these signs, and probably do not notice the conceptual structure of the exhibit areas of the museum. However, this structure is used for teaching by university and elementary-school teachers and Exploratorium staff teachers.

We have different expectations or goals in the use of the museum. We do not feel that an exhibit has 'failed' if a casual visitor misses the main point of an exhibit. But if that same exhibit fails to communicate a certain point to a group of teachers I am working with, then there is something wrong. Our visitors come for many different reasons, and in turn use the exhibits in many different ways, from very formal instruction to very informal sightseeing. Our expectations regarding the exhibits recognize this vast difference. A very good study was done in 1980 by Judy Diamond on the different ways families use the Exploratorium. This was her thesis for a doctor of philosophy degree (University of California, Berkeley) and it points out many very interesting teacher–student roles that family members take on at the various exhibits.

Falk: I would like to believe that the choice of perception was no accident. The topic touches people's lives in ways that make it easier to communicate science. Certain topics may be more appropriate for exhibition than others. It certainly helps if topics can be related to the individual.

Duensing: We are trying to exhibit some mathematics now. You can't teach people how to calculate in a museum but you can expose them to ideas of mathematics. We have numerous exhibits on exponential growth that give an idea of what this rate of change means. We also plan to have exhibits on randomness and prediction and non-linear rates of change.

Bell: If a student makes sense of one of those exhibits in his or her terms, what opportunities are there within the Exploratorium to explore those understandings or to have further questions answered?

Duensing: The exhibit sign tells you something about what is going on. Then we have 'explainers' in different areas of the museum whom visitors can ask for help or further information. There are also booklets or catalogues on several different areas of the museum, and short brochures called pathways for teachers to use with their students. These give more detailed information about the exhibit areas and guides to specific sections as well as related classroom activities.

Bell: Have you evaluated whether having those questions answered helps in the learning process?

Duensing: No.

Hearn: Essentially you provide an enjoyable learning process which probably takes a fair bit of time per person to appreciate and to use. How many people visit each year, how many make repeat visits, do you charge people to come in, and have you explored the idea of creating small scientific toys that could reach a wider audience?

Duensing: Just over half a million people a year visit the Exploratorium. We could get a few more but it does get very crowded at the weekends.

We did not charge for admission until about six years ago. It costs $4 for anyone aged 18 or over but is free for those younger than that. Children could often pay 50 cents or so but we don't want them to have to decide between coming in or buying a toy or a hamburger. The $4 ticket is valid for six months and in addition to encouraging repeat visits, this has psychological benefits.

The return rate varies between 10 and 12% of our adult visitors. We don't know how many of the children make return visits but many of the schoolchildren who come in at the weekends with their families have obviously been there before: we see them bringing their parents to their favourite exhibits. They see it as a place they want to share with their parents.

We have thought a lot about the marketing and whether toys would help. The store is an educational component in the museum but it makes money as well. We are just getting started on simple toys and we would actually like to do more.

Miles: Do you use science students or arts students as explainers?

Duensing: High-school students act as explainers for the general public. Some of them may be taking science classes but they are hired because of their interest in the place and how articulate they are. They are not usually asked to go deeply into explaining scientific phenomena or principles. It is more a matter of getting visitors engaged in another aspect of answering superficial questions. Sometimes a visiting physicist tries to challenge an explainer. When the explainers don't know the answer, they are encouraged to say 'I don't know. Let's see what we can figure out'. Some science students are terrible as explainers because they don't know when to stop.

We use older explainers to cope with school groups, as these can be difficult to handle. These explainers need to understand certain scientific concepts in more depth. They walk the kids through certain pathways.

Caravita: Some of the presentations seem to be so manipulated by the technology that the phenomena may appear far removed from everyday life. It may be difficult for people to bridge this gap.

Duensing: The 'so what' question I mentioned earlier is intended to help visitors over that gap. We built an exhibition with duplicates of 90 of our light and vision exhibits for the IBM gallery in New York City (the exhibition is now in the New York Hall of Science). In almost all the graphics we had a picture either of something from the outside that related to the phenomenon or of another interesting way to see the phenomenon.

Another example is with a poet-in-residence we had who saw that there were many connections between how people find meaning visually and how they find meaning in sounds and symbols. The National Science Foundation funded us to develop some exhibits on language. We use more computers and pieces of technology there than in any other area, yet a typical comment is 'I really like the language section because it is very non-technical, it presents all these non-technical ideas and I am not a very technical person'. It is interesting what people perceive and what is actually going on.

Caravita: What kind of exhibits do you have in the biology pathway?

Duensing: We still have a lot of work to do in biology. Many things are very small and the timing is tricky. Things are so fast or so slow that you can't really perceive what is happening during the time scale of a museum visit. We have a nice group on the nervous system and nerve responses and have added a new section on animal camouflage. We are just starting a major expansion in this area on membranes and cellular responses.

Science education through graphics at zoos

BEVERLY SERRELL

Serrell & Associates, 5429 South Dorchester, Chicago, IL 60615, USA

Abstract. As informal teaching devices, zoo graphics contribute to science literacy by helping people to formulate questions about nature and seek answers from the observation and interpretation of natural phenomena. By developing graphics which exemplify some of the principles of effective exhibit design in museums, zoos are successfully attracting the attention of visitors and promoting favourable behaviours such as reading, looking more closely at the exhibits, reading aloud to other members of the social group, and pointing at and discussing information about animals on display. Although these encounters are brief, the zoo's objectives of promoting visitor enjoyment, understanding and appreciation of animals may be enhanced.

1987 Communicating science to the public. Wiley, Chichester (Ciba Foundation Conference) p 147–160

A personal introduction

I got into the business of making zoo signs after working for the years 1970 to 1978 at the John G. Shedd Aquarium in Chicago, where I developed prog-rammes for visiting school groups and adult classes. After creating many activities for groups, I became more concerned about the kind of experience the average, unguided visitor was having in the galleries. Data gathered in a 1975 survey in which visitors were interviewed at the exits (Serrell 1977) and in a 1976 tracking and timing study of visitor behaviours in the galleries (Serrell 1978) led me to believe that people's enjoyment of the exhibits could be greatly enhanced if more educational aspects were added to the exhibits. It seemed that exhibit labels were often the weakest part of many museum and zoo exhibits, but they also seemed the easiest to change. I had also been influenced by the museum and zoo evaluation studies by Screven (1974a, 1974b, 1976), Shettel (1973), Rabb (1969, 1975) and others who spoke about the need for better educational exhibits. Since it wasn't my job to change the situation at the Aquarium, I went out on my own in 1979 to research the effectiveness of signs and try to improve them elsewhere, inspired by Roger

Miles' writings and struggles at the British Museum (Natural History). Brookfield Zoo and Field Museum of Natural History in Chicago were cooperative and supportive of my efforts and while I was doing this research I wrote a book about museum labels (Serrell 1983).

What is happening in zoos today?

Many American zoological parks are renovating old exhibits and building new ones. As sites for leisure-time recreation, zoos and aquariums have always been popular, but they are now enjoying new status and success, being hailed as urban economic growth industries. Aquariums, in particular, have been integrated into renovated waterfront malls in several American cities—Baltimore, Maryland; Boston, Massachusetts; and Monterey Bay, California—and at least six more are in the planning or building stages.

Using the word zoo to describe an out-of-control chaotic situation is becoming archaic. Instead of the old random 'postage-stamp' animal collection (i.e. one of each type), new aquarium and zoo exhibits are now planned with visitor traffic flowing through a well-orchestrated set of experiences and thematic encounters with animals and graphics (plus restaurants and shops). Educational graphics—a term that includes signs that are all text and signs with text and art on display with the live animal exhibits—are becoming an integral part of the display philosophy, along with more naturalistic exhibit techniques, e.g. fewer barred cages, more natural plants and rocks, and more mixed species. There seems to be more of a trend now to include the exhibit's educational story-line in the planning stages of new exhibits, rather than making graphics a decorative add-on.

In the past eight years, over a dozen zoos have instigated programmes of new graphics, and several of those have been funded by U.S. Federal grants from the Institute of Museum Services which has also recognized the importance of educational graphics. The American Association of Zoological Parks and Aquariums' self-accreditation programme of standards of good nutrition and comfortable housing for zoo animals also has standards for services for the zoo-goers, and the presentation of educational graphics is one requirement for accreditation, although no specifics are laid down as to what type or how many.

What do zoos teach?

Themes involving animal behaviour, adaptations, and a conservation ethic—encouraging the protection of endangered species—are commonly found in zoo programmes and graphics. Zoos see themselves as arks for conservation and have been putting time and money into breeding programmes for endangered animals, with some species actually intended for release back into the wild. As a public relations message, this story is a very positive counterpoint for those people who may think of zoos as animal prisons.

There is an obvious opportunity to teach the zoo-going public about zoology, for example about animal shape, size and colouring, along with the chance to communicate more complicated concepts of genetics and evolution (when the zoo's captive breeding programmes are discussed or when primates' hands are compared, for example).

Methods of teaching include special programmes led by zoo staff which are available at set times for school groups, demonstrations offered for general visitors, and supplemental materials available for teachers. The WIZE (Wildlife inquiry through zoo education) teacher materials developed at New York's Bronx Zoo are notable (Berkovits 1985). Many zoos have well-qualified educational staff (Bruggeman 1985), and museums are recognizing the creative and enthusiastic efforts by zoo educators.

Both general visitors and schoolchildren can benefit from zoo graphics that supplement the live exhibits. Unlike special programmes, which are only available at set times, graphics are always accessible. They optimally provide opportunities for what Lucas (1983) calls 'accidental encounters' for informal learning. Visitors do not typically go to zoos to study the displays or to learn from the graphics, but well-planned, intentional learning experiences— planted by the zoo staff—can be encountered by the unsuspecting visitors and before they know what is happening they can become involved in reading and making observations.

Is it science?

As informal teaching devices, zoo graphics offer science literacy by helping people to 'formulate questions about nature and seek answers from observation and interpretation of natural phenomena'—an educational outcome encouraged by the National Science Board. People come to zoos to look at the animals, and graphics can direct their attention to look more closely at the 'natural phenomena' on display. Graphic devices such as questions can raise people's curiosity, and interactive hinged lift-up labels can reinforce information gained by those seeking answers from the exhibits. Explanatory labels interpret natural behaviours commonly displayed by certain animals, such as mutual grooming, vocalizations or breeding. Motivation to use the graphic information, however, is often influenced by the level of animal activity. An awake animal has far more attracting and holding power and prompts more inquiry than a sleeping one.

Do the graphics work?

Characteristics of the graphics themselves—such as placement, legibility, readability and applicability to the exhibit at hand—also influence visitors' motivation to read and look. Graphics placed close to the animal discussed, written so that no prior specialized knowledge on the subject is required to understand them, in large typefaces, work better than ones placed behind the normal sightlines for the exhibits, in technical language, or printed in tiny type. (Common sense? Yes, but often harder to achieve than you would believe.)

FIG.1. On one side of a free-standing kiosk, 'Bat Facts' are presented in line art, photographs and large and small text. Milwaukee County Zoo.

FIG.2.   Animal identification labels using common and scientific names, a full-body silhouette, text, and an illustration of a special adaptive feature. Belle Isle Zoo, Detroit.

FIG.3.   Part of an exhibit of domestic animals, this panel has a tile-relief graphic (upper right-hand corner) of a primitive hunter, computer-generated art of the transformation of a wild cow to a dairy cow, and participatory elements: egg wheel and lift-up labels. Royal Oak Zoo, Detroit.

FIG.4.   Large panel next to tiger exhibit describes the international zoo breeding loan programme called the Species Survival Plan. Louisville Zoo.

By developing graphics which exemplify the principles of effective exhibit design in museum, such as brevity, concreteness and visual appeal, zoos are successfully attracting visitors' attention and promoting favourable behaviours such as reading, looking more closely at the exhibits, reading aloud to other members of the social group, and pointing at and discussing information about animals on display. Several in-house studies have shown that graphics can increase these desirable visitor behaviours. The gorilla graphics at San Francisco Zoo, the new animal identification signs at Philadelphia Zoo, and the key signs for lions and tigers at the National Zoo in Washington DC are good examples. Writers and designers of these and other signs have shared their secrets (Rudin 1979, Taylor 1981, Rand 1985).

These findings are similar to those found in museums. Borun (1980) reported that non-science-trained visitors learnt from labels in a science museum. She also found that more people read shorter labels more completely, similar to what others (M. Chambers, unpublished, Serrell 1981, 1982) saw in their label studies. Evaluation studies of labels which do not conform to the principles of concreteness, brevity and clarity have repeatedly shown that those labels fail to capture and hold attention and teach effectively (Munley 1982, Wolf 1981).

Beyond static graphics

While static graphics can be involving, overtly interactive devices are being added to some exhibits. Louisville Zoo's MetaZoo, Toledo Zoo's Diversity of Life, and San Francisco Zoo's Primate Center are all geared toward hands-on interactions to promote learning. Rosenfeld (1979) described some pioneering efforts at San Francisco Zoo's Animal Fair, and Jenkins (1985) has reviewed many current examples in zoos. In my opinion, the most intensive, involving and effective approach has been the National Zoo's Herplab—a family learning centre in the zoo's herpetarium. The Herplab has a specially staffed room with boxes of touchable items with directions for examining them, and provocative questions. HerpLab has been copied by other zoos, and the National Zoo provides a valuable model in their book *Families, Frogs, and Fun* (White & Barry 1984).

FIG.5.   Flamingo feather specimens displayed with art, text and photo on one side of a four-sided kiosk in the Aviary. Royal Oak Zoo, Detroit.

A personal conclusion

Although encounters between visitors, animals and graphics are typically brief and unstructured, the zoo's objectives of promoting visitor enjoyment, understanding and appreciation of animals may be enhanced. It is my hope that visitors will *not* see graphics as intentional learning devices, nor will they go to zoos seeking science information. My hope is that zoo graphics will be designed so that they subtly reinforce the visitors' visual experiences and that reading the graphics will be a painless and natural extension of viewing the exhibits.

If I have been successful in making scientific information accessible to zoo visitors—which is what I interpret Macdonald-Ross (1977) to mean by the term 'transformer' of information—then maybe an improvement will be made, away from the 'communicative incompetence' that Lewis (1980) points out as being so common. It is harder to do than I thought it was going to be when I wrote the book on labels in 1980. The typical power-broker structure of museum politics (Miles 1986) also applies to zoos. The results you see in Figs. 1–5 represent many constraints of budget and time and compromises of intention between curators, keepers, directors, designers and me. Nevertheless, I invite your feedback and critique.

References

Berkovits A 1985 Wildlife inquiry through zoo education project WIZE: an investment for the future. American Association of Zoological Parks and Aquariums [AAZPA][a] Annual Proceedings, p 369–375

Borun M 1980 What's in a name? A study of the effectiveness of explanatory labels in a science museum. Franklin Institute Science Museum, Philadelphia, PA

Breuggeman J 1985 The care and feeding of zoo educators. AAZPA Annual Proceedings p 362–365

Jenkins D 1985 A survey of interactive technologies. AAZPA Annual Proceedings p 72–79

Lewis B 1980 The museum as an educational facility. Museums Journal 80:151–156

Lucas AM 1983 Scientific literacy and informal learning. Studies in Science Education 10:1–36

Macdonald-Ross M 1977 Graphics in text. Review of Research in Education 5:49–85

Miles R 1986 Museum audiences. International Journal of Museum Management and Curatorship 5:73–80

Munley M 1982 Telltale tools: an experimental exhibition. Smithsonian Institution, Washington DC

Rabb G 1969 The unicorn experiment. Curator 12(4):257–262

Rabb G 1975 Signs—essential link with the public. AAZPA Annual Proceedings, p 140–143

Rosenfeld S 1979 The content of informal learning in zoos. Roundtable Reports 4(2):1–3, 15–16

Rudin E 1979 A sign for all seasons: from writer's clipboard to zoo exhibit. Curator 12(4):303–309

Screven C 1974a The measurement and facilitation of learning in the museum environment: an experimental analysis. Smithsonian Institution Press, Washington DC

Screven C 1974b Learning and exhibits—instructional design. Museum News 52:5

Screven C 1976 Exhibit evaluation—a goal referenced approach. Curator 19(4):271–290

Serrell B 1977 Survey of attitude and awareness at an aquarium. Curator 20(1):48–52

Serrell B 1978 Visitor observation studies at museums, zoos, and aquariums. AAZPA Annual Proceedings, p 220–233

Serrell B 1981 Researching visitor reactions to labels in a museum. AAZPA Regional Proceedings, p 261–275

Serrell B 1982 Zoo label study at Brookfield Zoo. International Zoo Yearbook 21:54–61

Serrell B 1983 Making exhibit labels: a step by step approach. American Association for State and Local History, Nashville, Tennessee, USA

Shettel H 1973 Exhibits: art form or educational medium? Museum News 51(1):32–41

Taylor L 1981 Gorilla gorilla gorilla gorilla gorraphics: an evaluation study of the graphics at San Francisco Zoo's Gorilla World. AAZPA Annual Proceedings, p 217–223

White J, Barry S 1984 Families, frogs, and fun. Office of Education, National Zoological Park, Smithsonian Institution, Washington, DC

Wolf R 1981 Hey, mom, that exhibit's alive: a study of visitor perceptions of the coral reef exhibit. Smithsonian Institution, Washington DC

[a] The address for the American Association of Zoological Parks and Aquariums is AAZPA, Oglebay Park, Wheeling, West Virginia 26003, USA.

DISCUSSION

Hearn: Why do people want to use a uniform style for labels?

Serrell: To simplify things so that they don't have to think so hard.

Laetsch: A uniform design is an article of faith with designers.

Miller: It seems to me that labels on exhibits in zoos and museums convey classification information to the public, rather than science or scientific literacy. In their older days, the Field Museum in Chicago had row after row of stuffed animals with little name labels beside them. Now they have put the animals into ecological types of exhibits, so visitors get some idea of how the animal exists in the world. In an exploratorium people are making discoveries for themselves and that can convey some sense of science.

Laetsch: Would you say that classification is not science?

Miller: By itself it is not science. Classification usually implies a grouping based on a theory or an underlying set of assumptions. It is not just a question of similarities and differences.

Hearn: There is not much difference between the public looking at graphics in a zoo and scientists going around a scientific exhibit or a poster session. We find that using big interesting pictures and very short captions is the only way to hold attention. Thirty to fifty seconds is the maximum time that even scientists give to such exhibits if they have a choice of many things to look at. When it comes to funding agencies and politicians, we have to grab attention, through interest, within five seconds or the individual moves on.

Laetsch: Poster sessions at scientific meetings often violate every rule of good communication. Presenters are only beginning to realize that people don't read small print and badly done graphics. People in zoos and museums have understood this for a long time. This kind of understanding will help the communication of science between scientists.

Gregory: Reading while standing up is incredibly tiring, though. Another reason why people spend so little time reading information on exhibits is that most captions are boring.

The interesting thing about classification is how it is done. But if you are assuming the theory of evolution when you classify species, the entire evolution argument is circular.

Some hint of underlying concepts would switch any reasonably intelligent person on. The idea of removing anything which is abstract or difficult in an attempt to make it more palatable is nonsense. Intelligent people like the feeling that there are problems, even in how to classify things. It might be useful to relate how you classify things in a supermarket to what happens in zoology. How far is classification related to a paradigm, or a theoretical way of looking at things, and how far is it simply because they look that way?

Deehan: The problem is essentially an advertising problem. First you have to attract the visitor's attention and then you have to sell something. What you are selling is information. I wouldn't look twice at a family tree but I would look at the rather beautiful line drawings Beverly Serrell showed us. Then you can start to tell people about bats, why they fly and so on.

Miles: The catch in what Richard Gregory said is the word intelligent. In The Dinosaurs and Their Living Relatives exhibition in the Natural History Museum we started the section on the concept of evolutionary relationships with some familiar organisms before going on to deal with dinosaurs. Evaluation showed that a sizeable minority of our visitors could not conceive how something that lives in the water could possibly be related to something that lives in the air or on the land. Communicating with the general public is different from communicating with a few intelligent people.

Gregory: I think my policy switches a lot of people on without switching anybody off, if it is done properly.

Hearn: Graphics and rapid reporting can tell intelligent people some facts which they can pursue if they want to do so. But are you suggesting that you can put across convergent versus parallel evolution through a simple analogy with classification in a supermarket? And would you use that analogy rather than using the animal itself?

Gregory: Yes. We think through analogies. If you think about joints and levers you want something about how the lever system can work in your limbs.

Lucas: There is a major problem in getting people to focus on the appropriate part of an analogy. I am not sure how you can overcome this.

Nelkin: Science is not information; it is concepts, ideas, theories, principles. No scientist would ever define science simply as information or facts. The whole exercise is fruitless if that is how you are defining science.

Hearn: One wants to get the public more interested in what the problems of science are. What are the benefits that might result from the new discoveries and how does scientific method approach these problems?

Nelkin: Tossing out lists of information or classification in itself is not interesting to most people, whether they are educated or not.

Miller: Every bus tour company in Washington, DC, tells you that if you spend one minute at each exhibit in the Smithsonian, it would take you 650 years to go through it. Most people would have difficulty in investing so much time. In recent years, unified theories in physics and other fields have appeared that provide a way of summarizing many areas. These grander themes may be lost under the weight of the trivia.

Screven: This discussion illustrates some of the things that often happen during the development of science exhibitions that almost guarantee that an exhibition, when installed, will not work as intended! We have here a group of bright people, highly knowledgeable about science, talking about how one should or should not go about teaching science—with exhibitions or whatever. There have been arguments over goals, over what should serve as content, what viewers may or may not find confusing, boring, useful, or exciting, how exhibits need to be designed, and so on. Apart from the deficits in visitors' knowledge of science, little has been said about the learning conditions that might be required if the target audience is to cooperate, or about what we need to know about visitor populations to match given goals with viewer needs, biases and interests, or about how we might empirically determine (via formative testing) how realistic our exhibit goals, content and exhibition methods are from observations of actual visitor responses.

This is what usually occurs in many planning sessions for science exhibitions—except that these are on a grander scale because they include not only subject experts but also administrators, donors, visual designers, and exhibit makers. This adds up to pandemonium, some of which we have seen here, and to design decisions that assure content accuracy, aesthetic beauty and good construction, but little or no science education of the kind intended. Poor science communication seldom matters, however, because the final exhibits usually meet the other major needs of their planners—the exhibits don't fall down, the text is scientifically sound, the aesthetics are grand, jazzy, and visually exciting, and the newspapers and trustees are happy—and, besides, the public seldom complains and even seems to enjoy it! As I have already said, the public is seldom if ever a part of this process except as a hypothetical entity in the heads of planners who speak of what visitors will or will not do, think or understand, supported by anecdotal remarks, personal experiences and references to attendance figures.

Serrell: A big part of science consists of making observations and zoos are appropriate places to encourage that behaviour.

The techniques of formative evaluation are useful. Before a sign or label is even put up you can fine-tune the information and the presentation to the point where you can be confident that most visitors will get your message. Then through experimentation you can see whether visitors are getting involved with the information. What they carry home with them is a hard question to follow. Staff at the National Zoo have studied the impact of their family learning centre. They have done telephone interviews with visitors two months or so later and they have some evidence that people do indeed follow up on information obtained in the centre. I am most concerned with seeing some *immediate* evidence that the information is being used, because if it is not being used it is not going anywhere.

Duensing: What happens later is something we would like to pursue too, but in terms of immediate evidence of effectiveness once we put an exhibit on the floor that is really just the beginning of the exhibit development process. Through observation and talking to people we try to find out what makes sense to them. We go on changing the graphics as well as the structure of exhibitions after they have been put out on the floor.

Hearn: Adults do not often visit zoos or museums with the idea of being taught. But they are perhaps using their children as intermediaries and then learning. Is that part of the public perception of science as learning? Do adults feel that they do not need to go to these places?

Lucas: The children may be the excuse for going. The other problem that Roger Miles was pointing out earlier is that the public perception or the mythology is that these are places for children. A lot of adults look at the children's programme *Blue Peter* on television. The people who actually use programmes may be quite different from the people they are intended for. It is not acceptable behaviour to go to the zoo by myself to learn something about the animals; it is much more of a social phenomenon to take the children.

Nelkin: One of the most pernicious aspects of the public conception of science, encouraged by a lot of museums, is the notion that science is a set of facts and that all facts are equally valid. The support of creationism in the United States, for example, is based on this notion. But science is much more than a set of facts, and facts are only meaningful within a theoretical and conceptual context. Anything that gets away from the notion that science consists of facts alone is positive. Anything that contributes to that idea is feeding the most naive and pernicious dimensions of the public understanding of science.

Tisdell: Some museums and so on don't even succeed in communicating at the lowest level—telling you what you are looking at. The next step, getting a few facts across to people, is useful too. At the third level some theory would be great but maybe we cannot achieve all the levels for everybody.

Friedman: Beverly Serrell showed us labels without the exhibits and Sally Duensing showed us exhibits without the labels. But labels are an important part of exhibits. The labels in the London Zoo are strikingly different from 10 years ago. They now explain why monkeys show big bare bottoms at certain times of the year. Presenting that as an isolated fact would give a false image of science as a list of curious facts but on seeing the label and the monkeys I could interpret the facts in a context. I could see that one monkey must be a female and another must be a female not on heat, or a male.

Falk: Theory without facts is not useful either. What separates a Darwinian view of evolution from a creationist view of evolution? If all you start with is a theory, both views are equally valid. It is presumably the evidence that makes you say that one theory makes more sense.

Nelkin: The difference lies in the nature of the theoretical constructs, not the facts.

Evered: Earlier Richard Whitley said that we should be talking about who is communicating what and to whom and what the consequences will be. But we keep shying away from how we evaluate what we are doing. How can we evaluate the impact of the various methods of communicating science to non-scientists that we have been talking about?

Whitley: I want to tackle the idea that there are facts which decide whether creationism or evolutionism is correct. Richard Gregory's work alone shows that the perception of facts is a complex process. Evidence is contextual. You can't immediately provide evidence for plate tectonics as opposed to other forms of geophysical change or whatever. The history of science is full of examples of where facts do not speak clearly.

Falk: I am suggesting that the two need to be dealt with in tandem.

Whitley: One person's facts are another person's theories.

Laetsch: This demonstrates that you can't understand science without understanding some history and philosophy.

Dixon: Darwin said he worked on Baconian principles, collecting facts without any sort of theory, but it is clear that he didn't do that. Some scientists, however, work in a mindless sort of way, collecting observations without any real guiding theory behind them. I think we are talking about good science versus bad science. Perhaps this is part of the variety of science.

Laetsch: You are right, but up to a particular point Darwin was not working with a theory. After a while it began to come together and he began to be less strictly inductive about things. He had to get the facts first, though.

Whitley: He still relied on a classificatory system. Any classificatory system implies some idea of order, of what the world is like.

Miller: I think the question David Evered raised is exactly on the topic of the conference. Are we talking about the communication of science, are we talking about the communication of science for purposes of literacy, or are we talking about the communication of scientific information? I don't think that communicating information *per se* includes literacy. We need to keep in mind the purpose of communication.

Laetsch: That comes back to the definition of literacy. There are differences in how we look at that.

Miller: We cannot expect everybody who walks through a museum to have the same facility with categorization as Darwin had. We must try hard to convey to people that part of scientific activity consists of building these models or having some models in mind when we try to organize facts in some fashion. We could try to give people walking through a series of exhibits the opportunity to see how science works. We could try to provide a unifying experience.

Tisdell: If a person didn't have any prior knowledge I doubt if they could pick up the whole thread of the theories. If someone has prior knowledge it is nice to see this reinforced. You have to cater for people with different degrees of knowledge. The *National Geographic* magazine does that, with a short broad description first and then more detail. Both aspects are important.

Lucas: As Dorothy Nelkin said, though, you have to be careful that you don't preclude the possibility of going further about the nature of the science. You must not limit visitors to the facts that can be found in an encyclopaedia of science.

Nelkin: I would go further and say you must not convey the idea that science is a collection of facts. That is worse than conveying nothing.

Laetsch: Zoos are the single most popular activity that we have. People go there to see the furry animals, not to learn theories of science. It is a social experience in many ways, but within that experience people can find things out. If they find there aren't many tigers left in the world, they may vote for conservation activities. This aspect of the zoo may come closer to meeting its objectives than providing science centres that exist to train people to help the economy or to vote in a more interesting way.

Miles: There seems to be a feeling that natural history museums are full of stuffed animals and nothing else. But any major natural history museum nowadays has exhibits that deal with the grand unifying theories of biology.

Caravita: Theories are implied in the information provided in museums and zoos. Scientists select the information they want to convey and there are theories behind the way animals are labelled or displayed according to their geographical regions. Probably it would be better to make these theories more explicit rather than leaving them to be inferred by visitors.

Science theatre: an effective interpretive technique in museums

SONDRA QUINN AND JACALYN BEDWORTH

The Science Museum of Minnesota, 30 East 10th Street, St Paul, Minnesota 55101, USA

Abstract. The theatre programme at the Science Museum of Minnesota is used as a case study (its history, goals, methods, and how it is integrated into the overall interpretive goals of the museum) to examine the use of theatre as a component of interpretive programmes. Theatrical forms of interpretation are described, as well as questions of attraction, entertainment, education and the capacity to engage many different kinds of visitors in science ideas and issues. A brief review is included of what is known about how people learn in the informal learning environment of museums and which factors encourage them to devote attention, time and effort to learning. Other museums are also using these techniques in their interpretive programmes. Some examples are given, as well as the results of the few formal evaluations that have been conducted to date. The Science Museum of Minnesota has concluded that education and entertainment are not mutually exclusive in any museum interpretive programme. Science theatre teaches science—facts, concepts, issues and attitudes—to museum visitors while entertaining them.

1987 Communicating science to the public. Wiley, Chichester (Ciba Foundation Conference) p 161–174

One of the most commonly (and fondly) held tenets among those of us who consider ourselves museum professionals is the idea that museums have immense potential as learning resources for people of all ages and backgrounds. Unfortunately, while we generally accept this theory, its proof has eluded most investigators. We need to learn more about how people learn in museums, how much, and what interpretive techniques make these settings more effective. During the past 15 years, The Science Museum of Minnesota has successfully used theatre as an interpretive technique to influence public perceptions of the scientific process or attitudes towards science and scientists. We believe theatre presentations are not a form superior to more traditional methods of interpretation. Rather, as we have discovered, it is a method very different from other interpretive methods, being both entertaining and educational. It will be some time before the potential of science theatre as an interpretive technique in museums is achieved. Implementation of research into both cognitive and affective learning is required.

Museums as informal learning environments

We know that there are several important differences between formal learning settings such as schools and informal settings such as museums. Museums as learning environments are characterized by free choice, the absence of prerequisites and credentials, heterogeneity of learner groups with respect to age, background and interests, and by the importance of social interaction as a criterion or ingredient of the visit (Laetsch 1979). In addition to their varied educational background and interests, museum visitors have diverse attitudes, preconceptions about what they see, and amounts of time they have available. They also have a variety of skills, knowledge and motivation. Unlike the required attendance of most traditional educational programmes, attendance at a museum is voluntary, and once inside, the course that the visitor follows is self-established. There is no 'required' reading, listening or viewing. Museum visitors may spend as much time as they choose at the things that interest them most. Kimche quotes Goodman as stating that people will learn best when they have opportunities to make choices about their own learning and chances to build on their own interests (Kimche 1978).

In many ways the museum is the antithesis of the traditional classroom. In the opinion of some critics, museums are one of the best forums available for public education. In an environment that is non-linear, self-paced, voluntary, and exploratory—qualities frequently lacking in schools—the ability to motivate and arouse curiosity is critical.

Since having a good time is a basic objective of most people visiting a museum, a natural corollary is that learning in a museum is recreation, for recreation means 'to create anew, to restore, refresh' (Tramposch 1981). And, much like most other recreational activities, attention and involvement depend on the quality of the experience and how interesting it is to the visitor. If an experience is boring or somehow unappealing, the visitor will move on. 'Like it or not, fun must be part of the experience. Fun is not incompatible with learning in the exhibit environment, but it must be used as a means to an end, not as an end in itself' (Screven 1986). The great challenge for museums is to link their communication objectives with the intrinsic exploratory, social and recreational interests of their visitors to create experiences that are both enjoyable and worthwhile.

In most museums, the physical structure of the exhibit *is* the exhibit. Until the early 1970s, the exhibits at The Science Museum of Minnesota were much the same—mostly static displays and dioramas. In giving serious thought to the problems of communicating scientific information with greater impact, the museum began to experiment with the use of theatrical techniques to interpret museum exhibits.

Initial presentations were simple monologues, researched, written, produced and performed by a single person on a platform set in the exhibit hall. These early theatrical presentations usually had an actor portraying a person from another culture, engaging audience participants in role play and explaining significant artifacts. These presentations were well received by museum visitors and were soon expanded to include more elaborate productions with wider dramatic scope.

After seven years of prototyping, the museum's theatrical venture had grown to become not only a unique and effective method of presentation but also part of the basic framework for the overall interpretive programme of the museum's new building. Theatre continues to thrive throughout the museum, existing in a variety of forms, from creative dramatics to formal productions. As many as 15 different pieces are performed at any one time in our six different exhibit halls, with the exhibits used as sets whenever possible.

The museum now employs a director for the programme and five professional actors who are part of the full-time staff. Playwrights, prop designer/fabricators, and costumers are regularly contracted for, while exhibit staff design and construct simple set pieces and create special lighting and technical effects.

Although an exhibit can give museum visitors a good introduction to the culture it represents or to the technology it demonstrates, it is sometimes difficult to communicate the full importance of an artifact, object or idea in two or three dimensions. The emotional impact, the all-important 'why did this come to be and what did it mean to the people who used it?', is difficult to explain. By bringing 'life' to an exhibit through theatre, we are better able to communicate an idea, evoke an emotional response and position ideas and objects in time and space. Through theatre, the audience can become closer to the creator of the object, the user of the object, the collector of the object, the purpose of the object and, most importantly, to themselves. The action and elements of theatre can explore the values, skills, experiences of real people or the happenings of a time. We believe that any object, person, culture or process can be presented dramatically, thereby increasing its educational potential for the museum visitor (Fig. 1).

What is science theatre?

Today, theatre is seen as entertainment whereas historically it was designed to educate as well as entertain. Theatre was once the vehicle through which societal morals, ethics, social issues and ideas could be communicated. Dramatic ritual served a number of purposes: the intention to influence events through 'sympathetic magic', to educate, glorify, entertain and give pleasure (Brockett 1968, p 3–4). Brockett concludes that 'although our specific goals may differ considerably from those of primitive man, the modern theatre still serves the last three purposes.'

As theatre has evolved in western society, it has become separate from education and synonymous with entertainment. Yet theatre is a very old, traditional and respected method of teaching and communicating ideas.

FIG.1.   Ka: Spirit of an Egyptian Mummy. Through the personification of one of the
spirits of the mummy, religious objects and beliefs and some of the spiritual history of
ancient Egypt are explained.

Today at The Science Museum of Minnesota, science theatre carries on these more traditional roles of theatre. For us, science theatre presents ideas and explains objects in an entertaining way to groups of visitors. It can (but doesn't necessarily) include sets, costumes, props, etc. 'The essence of science theatre is that we can do it without all the trappings of traditional theatre. It uses the exhibits; it is not on a set or behind the traditional fourth wall. Each of our dramatic presentations is adapted to each audience's level of understanding; flexibility is critical. The non-captive, transitory nature of the audience is also very important. Our visitors have other reasons for being in the museum than to watch theatre. We believe that most (if not all) of our pieces should involve the audience in some way . . . in ways that are very rare in traditional theatre, including the actors speaking directly to the audience as individual human beings' (T. Bridal, personal conversation, 1986).

A colleague who attended a recent workshop on using theatre as an interpretive technique in museums wrote, 'The surprise for me was the interactive nature of . . . [the] program. Watching the actors "play" to the audience—involving the children and adults in the same piece at different levels of understanding—was a revelation. The pieces are entertaining, but the opportunity for education is far richer than I would have guessed. The technique allows for more than emotional involvement for audiences—it engages them actively at their own level to enhance their learning about the subjects presented. Success depends on the actors' abilities to gauge an audience—to personalize the presentation' (P. Davis, personal communication, 1986)

The Science Museum of Minnesota draws on a spectrum of techniques to create significant learning experiences for visitors. These techniques range from more traditional interactive and static exhibits, including photographs, written text, computer games, lighting and audio effects, to interpretive programmes that include informal activities, demonstrations, theatre presensations and special events. Our goal, rooted in our philosophy of providing a participatory experience, is to transform all the museum into a 'stage' where the visitors can take an active role in creating their own experience.

As a concrete example of how the various elements of interpretive programming are actually used in an exhibit, visitors to the opening weekend of the exhibit, 'After the Buffalo Were Gone', participated in some of the following activities:

- Traditional Blackfeet Indian stick games;
- Storytelling about the realities of Blackfeet Indian life 100 years ago;
- An enactment of the confrontations between Almost A Dog, former Blackfeet Indian warrior and buffalo hunter, and Major Young, American Indian agent;
- A portrayal by an American Indian actor about the religious rights and implications of Chief Mountain, the sacred ground of the Blackfeet Indian.

'As an interpretive technique, theatre offers one of the most powerful ways

FIG. 2.   Diego de Landa. Lorenzo Cocon takes the crucifix from Father Diego de Landa and renounces the gods of the ancient Mayan people.

to present ideas and feelings in context. The very nature of theatre is to tell the story to personalize, through characters and plot, those ideas and feelings that might otherwise escape the visitor's attention. In the field of social anthropology, for example, theatre is an excellent medium for dealing with native legends, belief systems, and episodes of cultural conflict' (L.B. Casagrande, unpublished work, 1986). In one of three life histories featured in the exhibit, 'Cenote of Sacrifice: Sacrificial Wells of Chichen Itza', a historical perspective on the conflict between the religious beliefs of the Roman Catholic Church and the Mayan people is presented (Fig. 2).

'Theatre is fiction to which the audience consents, and thus it can provide a non-threatening framework for dealing with complex crucial social issues relating to science. Theatre makes it safe to contemplate disturbing controversies and to probe attitudes in science' (Grinnell 1979). In our piece 'Nuclear Winter', for example, issues relating to the possible aftermath of a nuclear attack are debated in a traditional town meeting. In a confrontation between a government representative and an anti-nuclear activist, strategies and plans for civil defence during a nuclear war are examined.

Difficult concepts or discoveries can be communicated through interpretation of how a scientist 'thought' or 'acted' on a problem, or how some specific aspect of technology affects people's lives, or how social, economic, geographic or technological problems can be solved. In 'Bring Back My Sweet Pea to Me', talking vegetables present young audiences with a problem and encourage them to think through the process of deciding how to transport vegetable produce to market efficiently (Fig. 3).

While exhibits help visitors to understand scientific facts, principles and concepts, drama also lets them experience the conflicts, challenges and adventures involved in scientific discovery and research. Based on a true story of scientists at 3M corporation, 'Space Crystals' introduces a physicist and a chemist who helped to develop the process of growing crystals in space aboard the space shuttle. This dramatic production, with an exhibit featuring actual crystals grown in space and 'The Dream is Alive', an Omnimax film about the space shuttle, formed a trilogy of experiences from which visitors could learn about contemporary space science.

A dramatized demonstration such as 'Games Around the World', brings a particular culture or aspect of a culture to life. The playing of games, for example, is a way to develop important skills among various societies, such as eye–hand coordination in hunters and gatherers, or the strengthening of family bonds or community relationships among sedentary farmers. Visitors can relate easily to games of other societies quite different from their own. In demonstrations of the games or dancing or even music of other cultures, visitors can examine various value systems, traditions and beliefs in a non-threatened way (Fig. 4).

Assessing the effectiveness of science theatre

Dramatic presentations are not an entirely new concept in a museum setting, yet few such programmes have been closely analysed, written about, or evaluated. As Laetsch states, a fuller exploitation of the museum as a free-choice learning environment depends on a greater understanding of the

FIG. 3.   Bring Back My Sweet Pea to Me. Rudy, a Minnesota rutabaga, encourages children to help him decide on the best way for vegetables to travel to market.

factors involved in eliciting curiosity and in motivating people to attend to an activity long enough for the educationally significant goals of the activity to be realized (Laetsch 1979). We simply do not know enough about how or what people learn in museums. We are ignorant of most of the factors that encourage or enable learning in museums, and we have few precise data on the effectiveness of dramatic presentations in science museums.

Professional opinions and anecdotal evidence are both common. Our interviews with professional staff from museums across the country as well as with many of our visitors produce much the same comments—that 'you can learn things from science theatre, even though you're having a good time'. Overall, the reaction of visitors to such performances is universally positive. People use words such as 'wonderful', 'enjoyable', 'just perfect', and 'beautifully done' to express their pleasure with the programme. The pleasure of the visitors is directly attributable to their finding a live performance in the museum. The acting is seen as a 'change of pace' by many and a way to better enjoy the museum by most (Munley 1982).

Without having conducted a formal study, the Museum of Science, Boston, can give some powerful evidence of the effectiveness of their production of a dramatic presentation called 'Dragon Bones'. A museum-sponsored poster

FIG. 4. Games Around the World. Actors demonstrate childhood games used as teaching techniques by various cultures around the world.

contest drew entries clearly inspired by the play, as shown by depictions of characters from the play, the set, and special effects used in the piece. Another indicator of the play's effect was the large number of unsolicited letters received from students, describing specific situations encountered in the play. Finally, children who had just seen the play were observed locating and identifying particular artifacts important to the plot of the play yet displayed quite inconspicuously in the exhibit (M. Dassault, personal communication, 1986).

The Franklin Institute, Philadelphia, can also report preliminary conclusions from an informal survey they conducted of an experimental theatre production, 'Wondrous Visions: A Visit with Leonardo da Vinci'. Observations of the performances revealed that they drew large crowds of viewers who remained throughout. Moreover, staff reported that the presence of the actor provided an opportunity for interested viewers to ask detailed questions at the end of the performance. In interviews with adults and children, even young children responded that they thought the play was fun and that they thought they learnt 'some things' (S. Garred, personal communication, 1986).

At The Science Museum of Minnesota, a survey conducted for the presentation of 'Sara the Scientist' showed that all but one respondent felt that the play enhanced their enjoyment of the exhibit, 'Women in Science and Engineering'. A significant number of visitors stated that their perceptions about women had been affected by seeing the theatre piece. Although all our studies to date have been informal we are now planning a more comprehensive programme of evaluation.

The most comprehensive study so far of the effectiveness of theatre as an interpretive technique was done by Munley in 1982. Her evaluation was designed to study the effectiveness of a dramatic presentation staged in a period-room exhibit in the National Museum of American History, Smithsonian Institution (Washington, DC).

The reactions of visitors to the performance were more intense than any the researcher had seen in any museum programme. People were involved in the relationship between the two characters and demonstrated enormous empathic abilities in their responses to what they saw and heard. In addition, Munley reported that visitors rated the dramatic performance as one of the best features of the museum.

Visitors saw the drama programme as significantly more active and involving than the traditional presentation of the period room. They reported a significantly more powerful and moving experience in the museum than did visitors who experienced the parlour as a traditional period display.

The Smithsonian's experiment in dramatic interpretation left a distinctive mark on the visitors who participated in it. It was successful in meeting the objectives of providing an involving experience for visitors. It also proved to be an effective vehicle for helping visitors to make connections between their lives and the lives of people who lived in other times. In Munley's words, 'Live interpretation is an enormously successful method for getting visitors involved in the museum experience. Visitors are unanimous in their requests for more of this programming in the museum.'

Munley's study provides much-needed information about visitor reactions to theatre presentations performed in an exhibit area and it presents information which, in her words, 'offers a first stab at coming to understand the unique contributions of differing interpretive methods.'

Conclusion

In conclusion, effective communication with a wide variety of museum visitors demands the use of a variety of techniques. Science theatre is only one of the techniques on an interpretive spectrum. To use it most effectively we need to look carefully at our educational message and decide if it is the most appropriate technique. We must set up economical evaluation techniques to pinpoint our successes and failures. We must be prepared to give up some of our comfortable habits—our conventional, traditional ways of thinking about education, in which we attempt to impose a formal structure on an informal setting. We must be willing to commit some of our already over-extended resources to experimenting with new methods. We can no longer afford to continue using only traditional formal educational methods in an informal learning environment. We must look 'outside' the realm of traditional methods and borrow from other fields of endeavour, such as theatre, to adapt and make those techniques our own.

[This presentation included extracts from The Science Museum of Minnesota's productions of 'Madame Curie' and 'Bring Back My Sweet Pea to Me', acted by Valerie Smith, graduate of East 15 Acting School, Loughton, Essex (Director, Margaret Bury).]

Acknowledgements

We wish to acknowledge contributions by: The Science Museum of Minnesota Staff— Tessa Bridal, Director of Theatre Program; Thomas Baerwald, Director of Geography; Louis B. Casagrande, Head, Department of Anthropology; Valerie Hamilton, Administrative Assistant; Joan Lisi, Actress/Interpreter; Orrin C. Shane III, Curator of Anthropology for Archaeology; and Phillip Davis, Pearson Museum, Southern Illinois University, Springfield, Illinois; Mary Dussault, Museum of Science, Boston, Massachusetts; Sheila Garred, The Franklin Institute Science Museum, Philadelphia, Pennsylvania; and James W. Quinn, President, K.Q. Associates, Minneapolis, Minnesota.

References

Brockett OG 1968 History of the theater. Allyn and Bacon, Boston

Grinnell S 1979 A stage for science: dramatic techniques at science-technology centers. Association of Science and Technology Centers, Washington, DC

Kimche L 1978 Science centers: a potential for learning. Science 199:270–273

Laetsch W 1979 Conservation and communication: a tale of two cultures. Southeastern Museums Conference Journal, p 1–8

Munley M 1982 Evaluation study report: buyin' freedom. National Museum of American History, Smithsonian Institution, Washington, DC

Screven CG 1986 Exhibition and information centers some principles and approaches. Curator 29(2):109–137

Tramposch B 1981 On interpretation. The Colonial Williamsburgh Interpreter 2(2):1–2

DISCUSSION

Tisdell: Science theatre must be very expensive.

Quinn: With an annual theatre budget of about $80 000, we reached 120 000–150 000 visitors, which is more than 20% of our total number of visits for 1986. Most of the shows we develop cost between $300 and $500 to write. Once we have developed a script, we can use it for a long time. Performance costs constitute the largest part of our budget. We have five actor-interpreters on the staff who are paid about $10 000 per year for a 40-hour week. They not only perform but also serve as interpreters for the exhibits. They also help other interpretive staff to develop their interpretive techniques.

Tisdell: Your staff are actively involved in this interpretive role while in other places they have a more passive custodial role.

Quinn: Other museums in the USA have been working with us to develop a similar programme. They have tried to train some of their interpretive staff to do some of the theatrical work but they have also hired actors to work as exhibit explainers part-time.

Tan: We cannot get our staff at the Singapore Science Centre to become actors on a regular basis because they have other functions apart from interpreting the exhibits. Instead we have a full-dress science show on a particular topic every two years or so. The staff rehearse for about a month and the 30–40-minute show is somewhat on the lines of the Molecule Theatre here in London. We get audience participation but of course the presentation is not continuous and therefore we do not reach as many people as you do in Minnesota. To make up for that, we distribute videotapes of the shows to libraries which can lease them to schools and community groups. The topics usually tie in with some campaign by society or by the government, such as environmental matters.

Quinn: It has taken us quite a few years to build our theatre progamme and to make theatre presentations an integral part of our interpretive programme. We have tried a variety of methods through the years, including the establishment of internship programmes through the university drama departments and working cooperatively with a local community theatre. Museums that do not have staff trained in the creative arts can also develop projects such as ours, using the methods described above.

Laetsch: Richard Eakin, a professor of zoology at Berkeley who is also an amateur actor, kept his students awake by dressing up and impersonating Mendel and Darwin and other scientists once a semester. He became so popular that there was standing room only at these events. He went on tour and on television. The students may not have got better marks but they certainly remembered that part of the course. However he said he wanted to be remembered as a scientist, not a ham actor.

Wolpert: An actor once gave a lecture on brain and behaviour at the AAAS meeting. It was just garbage but he presented it so effectively that he was rated very highly. In a study of students who had rated their mathematics teachers, the students' results were inversely related to the ratings. I have a lecture called 'Good lectures are bad for you'.

Friedman: Teachers who are more demanding and are the toughest graders are often rated more highly than teachers who let students get by without much work. If that hypothesis is true, one would expect the teachers who are rated as most professional to have students with the lowest averages.

Bell: In psychotherapy children may feel more able to speak to actors dressed as carrots, for example, because they are not threatening. Is there room for interaction in your science drama, and do the children speak more to the actors than to their teachers?

Quinn: Usually there is some interaction at the end of the piece. Some pieces are totally participatory, either verbally or non-verbally.

Hearn: This use of theatre is very intriguing but perhaps it is too complex and expensive to use widely. Have you analysed whether the visitors who stop to look are the scientifically literate ones?

Quinn: We don't know enough about the audiences at our theatre programmes. We have a membership base of 29 000 households so we get a lot of repeat visits from these members as well as casual visitors who encounter a single performance by accident. We are now working on a museum-wide visitor survey programme to further develop our visitor profile. This will include more accurate information about who attends our science theatre presentations.

Caravita: Some of the performances are about the history of science. Are the scripts based on actual documents?

Quinn: We try to use primary documents to develop scripts.

Caravita: Have any of the teachers tried to export this technique to their own classrooms?

Quinn: We have recently begun to send our science theatre productions out to schools, especially to those who are unable to visit the museum. A fee is charged which covers all expenses and reservations need to be made far enough in advance so that we can book actors and make all the necessary travel and scheduling arrangements.

The role of science books for the public

MICHAEL MACDONALD-ROSS

Institute of Educational Technology, Open University, Walton Hall, Milton Keynes, MK7 6AA, UK

Abstract. During the past two centuries science publishing has undergone drastic changes. The growth of the middle classes, the spread of literacy and the development of a mass market for books are three trends which should have produced a huge potential for works of science aimed at the public. In practice, professionalization has split working scientists from the general educated public, who now get their meagre understanding of science mainly from television and newspapers. This is a serious problem for the profession of science, whose funds come largely from taxation, and whose public image has been deteriorating.

The published book now sits uneasily between the research report in a journal and popularization in the mass media. Few books on science reach the fast-seller lists. Those that do are spin-offs from television series or books on personal or social topics which use science as a source of knowledge. In a market-oriented society people read books that have been sold to them as containing benefits they recognize and want. This means that scientists cannot determine in any *simple* way to propagandize their work and ideas. Such an effort would be likely to fail except where the product met a perceived need.

Certain publishing ideas may be helpful to scientists wishing to communicate with the public. The first step is to identify a clear need amongst well-defined groups of readers; the next steps are to use modern design, production methods and marketing in a professional manner. However, if the problem is how to allow the public access to scientific information as and when they need it, even this marketing approach will not by itself be sufficient.

1987 Communicating science to the public. Wiley, Chichester (Ciba Foundation Conference) p 175–189

There are intimate connections between the practice of science and the art of printing; science is a social enterprise in which the recording and transmission of methods and results is of supreme importance. The technology of print has met these needs quite well, for reasons which are fairly clear. For a start, the effectiveness of the printing process and the flexibility of the distribution system of publishers, booksellers and libraries must be important factors (see Febvre & Martin 1976 for an excellent account of the history of print).

Science does, of course, have some special communication needs to which print is well suited. William Ivins Jr., who was Curator of Prints at the Metropolitan Museum of Art, believed that the key advantage of print over hand-written or spoken communication was the ability of the printing press to reproduce illustrations and diagrams:

> The printing of pictures . . . brought a completely new thing into existence—it made possible for the first time pictorial statements of a kind that could be exactly reproduced during the effective life of the printing surface. This exact repetition of pictorial statements has had incalculable effects upon knowledge and thought, upon science and technology. (Ivins 1953, p 2–3).

Just how intimate the connections are between science and printing can be seen from this account by Kuhn:

> When the individual scientist can take a paradigm for granted, he need no longer, in his major works, attempt to build his field anew . . . that can be left to the writer of textbooks. Given a textbook, however, the creative scientist can begin his research where it leaves off . . . No longer will his researches usually be embodied in books addressed, like Franklin's *Experiments* or Darwin's *Origin*, to anyone who might be interested . . . Instead they will usually appear as brief articles addressed only to professional colleagues . . . only in those fields that still retain the book . . . as a vehicle for research communications are the lines of professionalism still so loosely drawn that the layman may hope to follow progress by reading the practitioners' original reports. (Kuhn 1962, p 19–20.)

According to Kuhn, then, once a normal science has matured in the sense of sharing a *paradigm* its use of print becomes specialized, as follows:

(1) Research communiques largely restricted to brief articles addressed only to professional colleagues whose knowledge of the shared paradigm can be assumed, and published in journals or conference proceedings. (Kuhn, p 20.)

(2) Textbooks, which are pedagogic vehicles for the perpetuation of normal science, and which communicate the vocabulary and syntax of the modern scientific language. 'To an extent unprecedented in other fields, both the layman's and the practitioner's knowledge of science is based on textbooks and a few other types of literature derived from them'. (p 137.)

(3) Popularizations, based or modelled on the textbooks, but using language closer to that of everyday life. (p 136.)

(4) Philosophical works modelled on the textbooks. (p 136.)

(5) Retrospective reflections upon one aspect or another of the scientific life. (p 20.)

So, to summarize, in the early stages of a science, books written by creative scientists serve several audiences, but when professionalism is established, and colleagues share a common paradigm, then published accounts become specialized, with different products serving different audiences. This is a fairly plausible account, which we shall have the opportunity of testing against the examples I shall present later.

Readership of science books for the public

Survey research data collected for the book industry can answer some questions. In the first place, book reading by adults is by no means a dwindling pastime. In fact, people spend relatively more on books than they did 10 years ago (Table 1).

TABLE 1   Consumer expenditure in 1985 at constant 1975 prices (from Curwen 1986)

Books	150
Magazines	80
Newspapers	83
Cinema	40
Other entertainments	161
Miscellaneous recreations	182
Radio/TV	238
Total consumer expenditure	125

The percentage of the Euromonitor sample reading a book at the time of the survey was 45% in the UK, a figure that has remained roughly the same for some years. When the sample is split into survey research categories based on occupations, ABs are most likely to be reading a book (65%) and DEs least likely (40%) (category A: higher managerial/professional; B: middle managerial/professional; C1: junior managerial, clerical; C2: skilled manual; D: semi-skilled manual; E: pensioners, unemployed). Sources of books are as shown in Table 2.

TABLE 2   Sources of books read by Euromonitor sample in the UK (Euromonitor 1985)

Source	%
Bought	35
Library	34
Borrowed	15
Gift	9
Other	7

Unfortunately, we cannot get detailed information about science books for the public from these surveys. There is no category that covers these books satisfactorily, and in so far as they can be identified in the surveys their role is insignificant in comparison to other kinds of books. For example, in recent years the category 'science, general' has had about 50–100 new titles per annum in the UK, where the total number of new titles per annum is over 40 000.

Scrutiny of the 'fast sellers' lists does, however, produce some interesting items. In 1982 the only book in the top 10 fast-selling paperbacks with science content of a sort was the F Plan Diet (1.5 million copies in the UK). In 1983 the top nine in this list were fiction and the 10th was on astrology! In the 1984 hardback non-fiction list there were two science books for the general public, but both were spin-offs from successful TV series. In the first place was David Attenborough's *The Living Planet* (BBC/Collins) which sold 125 000 copies at £12.00 and in eighth place was Jonathan Miller and David Pelham's *The Human Body* (Cape) which sold 27 000 copies at £9.75. The conclusion I draw from this and other evidence is that science books for the public have only limited commercial success unless they are spin-offs from successful television series, or provide the basis for practical or social topics of great interest to the general reader.

There is more to be said on the subject of readership for science books. We cannot make do, I'm afraid, with a simple dichotomy between 'professional' and 'public'. If by 'public' we mean literate non-specialists, that still leaves us with a wide range of people. They might be university graduates; they might be scientists in other specialties. They might, on the other hand, be early school-leavers with low educational attainment. Table 3 shows a classification, based on educational qualifications, which may be useful for readership research.

As you can see, Table 3 expands the census categories (mostly B and C1) to produce groups which are more relevant to our discussion. In this table all the groups from S2 to O are covered by the rather vague term 'public', yet they are vastly different in reading skills, knowledge of science and interests. Really, anyone from a cabinet minister to a dustman *might* in principle be a potential reader for a suitably designed science book but there is no way that any one book could serve such an incoherent audience. Few science books are read by category O and most, one would guess (there appear to be no survey data), are read by other scientists. I suggest, in the hope of provoking some survey research, that science books supposedly written for the general public are mainly read by teachers of science and fellow scientists who wish to keep in touch. To reach a wider audience books have to attack subjects the public is interested in, or be connected with a popular television series.

These days most people are educated—or at least spend a lot of time in scholarly institutions. That is a big change from the 18th century. Also notice how many readers have some kind of scientific or technical education: perhaps 20% of the population in a western country. Readers with a tertiary education are important in any industrialized country. They hold influential positions, they are the most likely groups to appreciate non-technical accounts of science, they influence the relative standing of science as a profession, and hence influence the funding of science. They are in a position to influence the young as to whether science is a worthwhile career, and it is their children, by and large, who form the mass of entrants into science courses at university.

TABLE 3 Classification of readers by type of education

Code	Readers based on education	Comments
S1	Specialists with same background as the author	Not part of the general public
S2	Scientists from other specialties	
S3	Science teachers	
T	Technologists	
P	Other professional qualifications	
H	Humanities graduates	
A	18+ school-leavers	
O	16+ school-leavers	
I	Illiterate members of the public	Not book readers, by definition. Functional illiteracy in US and UK about 20% of the population

By contrast, the less well educated groups are less knowledgeable about science, and their children are less likely, on average, to take up science as a career.

Examples

Table 4 offers some suggestive glimpses into the history of science books for the public. We see that the kind of science which the public is willing to read has changed greatly in recent years. Obviously changes in society and in the practice of science have altered the kind of books which people will read. One of our problems in thinking about these changes is that key terms like *science* are rather vague and need some definition. Do we interpret science narrowly, in the Anglo-Saxon tradition, to mean the physical and biological sciences, or broadly (as in the term *Wissenschaft*) to include all kinds of exact knowledge? Are we prepared to include the social sciences? Or technology? It makes a great difference to the conclusions we can draw. In Table 4 I have included not only science books in the narrow sense but also books based on scientific knowledge or scientific method, believing (as I do) that it is often through these channels that science reaches the general reader.

We see from the examples in Table 4 that there have been great changes in science publishing in the last two centuries. Of course, our society has changed a lot during this time, largely as a result of the scientific and technological revolution, and the practice of science has also changed greatly.

TABLE 4   Examples of books with scientific content read by the public

Author	Title, date and place of publication	Comments
Agricola G	de re Metallica, 1656	Woodcut illustrations of mining technology
Baker H	Employment for the microscope. London 1753	Copperplate illustrations separated from text. Mixed readership of amateurs and professionals
Simpson P	On the improved beet root as winter food for cattle. London 1814	Diffusion of practical agricultural knowledge
Squarey C	A popular treatise on agricultural chemistry: intended for the use of the practical farmer. London 1842	Diffusion of *scientific* agricultural knowledge from research by Davy, Liebig
Balmain WH	Lessons on chemistry. London 1844	Standard textbooks mark the maturity of a science
Mantell G	Petrifactions and their teachings. London 1851	Wood engravings. A specialized museum reference work, bought and read by amateurs and professionals
Darwin C	On the origins of species by natural selection. London 1859	Perhaps the last landmark work of science to be widely read by the educated public
Pepper JH	The boy's playbook of science. London 1862	Wood engravings. Aimed at the younger reader for use at home as a hobby. Experimental approach
Swinbourne A	Picture logic. London 1875	Use of humour and illustration in support of exact knowledge
Thompson CB	Scientific management: a collection of the main articles describing the Taylor system of management. Cambridge, Mass. 1914	Diffusion of scientific *method* into industry. Forerunner of operational research and systems analysis
Hogben L	Science for the citizen. London 1938	A genuine best-selling science book for the public: the last before the modern television era?
Bernal JD	The social function of science. London 1939	Pioneer work on the science and society theme
Allee WC	The social life of animals. London 1941	Example of modern work of science which is accessible to the public. Animal behaviour was a primitive science at that time

Hayek FA	The road to serfdom. London 1944	Gap between public and professionals less noticeable in the social sciences
Shannon CE & Weaver W	The mathematical theory of communication. Urbana, Ill. 1949	First part of the book addressed to electrical engineers, physicists and mathematicians, the second part addressed to the general reader
Huxley J	Soviet genetics and world science. London 1949	Interaction of science and politics
Eiseley L	Darwin's century. Evolution and the men who discovered it. London 1959	History of science as a means of reaching a general audience
Kahn H	On thermonuclear war. Princeton 1960	Use of systems analysis in military strategy
Kahn H & Wiener AJ	The year 2000. New York 1967	Use of systems analysis in long-term forecasting and planning
Meadows DH & DL, Randers J & Behrens W	The limits to growth. New York 1972	Use of computing allied to science and technology to connect economics with ecology
Feinberg G	What is the world made of? New York 1978	Books of this kind rarely succeed in reaching a general audience. Perhaps read by school science teachers
Hearnshaw LS	Cyril Burt, psychologist. London 1979	Biography as a means of reaching a general audience
Pearson D & Shaw S	The life extension weight loss program. New York 1986	Dietary information based on research

Trends in science and society

The following trends and changes are especially relevant.

(1) The professionalization of science, with the growth of disciplines and specialties within disciplines. In the 18th century a scientist might fairly be called a 'natural philosopher', but the 19th century saw the growth of professionalism. The term *scientist* was suggested by Whewell at the British Association meeting in 1833. Courses in science were offered in universities, textbooks were written and the modern disciplines were born. Darwin, for example, described himself as a *geologist*—and that despite his being a real natural historian in the old-fashioned sense of the term. A cause and consequence of this professionalism was the growth of a genuine scientific method. The procedures, methods, apparatus, laboratories, terms, theories, data, courses and qualifications had a catalytic effect on the growth of the scientific disciplines, but this professionalism also had the effect of splitting the specialist from the rest of the educated community.

(2) As a result of this process come the increased centralization, mechanization and cost of scientific research. So, at the same time as a science becomes less and less intelligible, it becomes more and more costly, centralized and bureaucratic. Also as a result of this process the amateur scientist virtually disappeared and there was a growth of the scientific 'priesthood', who became guardians of rational knowledge.

(3) The growth of technology is, from the point of view of the public at large, of more immediate importance than the growth of science. The two do not necessarily go hand in hand, but in fact they have done so in modern times.

(4) The growth of the middle classes in the early 19th century is important to our story. The main cause of this growth was the success of capitalism and the use of machine technology in industry. Scientists were then, and are now, members of the middle classes, and so for the most part are those who read their work and pay for it, or at least make the decisions about funding.

(5) The spread of literacy was one of the consequences of the growth of the commercial middle classes. A mass market for books was established between 1840 and 1890. The enabling factors were the spread of literacy and improvements in printing technology that brought the cost of books to within the purchasing power of the middle classes.

(6) The growth of state education led to the gradual decline of the self-improvement movement. This meant that scientists could only get access to the education of the young through the state. This increased the importance of published books, then as now almost the only uncontrolled channel of communication to the public.

(7) The 20th century has seen a great flowering of the social sciences, the roots of which growth lie in the 19th century. By and large the social sciences have not had the same difficulty communicating to the public as have the natural sciences. This is partly because the social sciences are not so specialized as the natural sciences (in other words, the process of professionalization is not so far advanced), and partly because their subject-matter is closer to the concerns of the public.

(8) Advertising is well established in the 19th century, and modern marketing methods developed in the 20th century. Fast-selling books are now always accompanied by, and to some extent created by, sophisticated marketing programmes.

(9) The 20th century sees science applied to war and also, partly as a consequence, increased scepticism and hostility to science. This hostility is now a serious obstacle to public appreciation of, and support for, the physical sciences.

(10) There has, in general, been a great and still increasing concern with the many interactions between science, technology and society. Some recent books which tackle this area have achieved fast-selling status, though in such cases the science content is just a means to an end.

(11) The advent of television as the primary mass medium is important. Of course, this is the medium through which most people get their knowledge of science. As explained above, the only *pure* science books for the public to reach fast-selling status in recent years have been spin-offs from popular television series.

(12) There is at present great interest in ways and means to improve the quality of individual life: ecology, health and fitness. Some recent books in this area have had huge sales and great influence—but again the science content is not the primary reason for purchasing or reading.

Modern publishing methods

To be commercially successful a modern science book for the public has to meet some of these criteria. If a book is about pure science, it should be connected with a television series, or it should be aimed at a coherent target audience that can be identified and reached by marketing techniques. School science teachers are one such target audience. If it is about applied science, the book should be on a topic of pressing social and/or personal concern for the reader. Again, such a book must be supported by a properly designed marketing programme.

Book production methods are changing under these marketing pressures. There is a growing tendency to pre-sell books in various languages before they are written. Once orders are taken, the book is produced by a writing and design *team*, rather as we do at the Open University. This approach is capital-intensive, but quite effective commercially, and it is potentially well suited to science books (assuming they meet a clear need).

Other production and marketing methods are possible in principle, but are almost unexplored in science publishing. For example, much publishing in the fields of law, taxation and investment is *subscription* publishing. This might suit certain areas of science such as atomic energy, where the public interest and need for information is long-term and highly committed. The most radical approach of all is to regard science and medicine as public properties, and ask the question 'suppose members of the public wish to find out about various topics, what kind of system can we devise that will allow them to do so?' This takes the emphasis away from *dissemination* of information and instead puts the emphasis onto *selection* and *access*. There is no good example that one can point to at present; nevertheless, this may be the most important problem to be solved in science communication today.

Summary and conclusion

At present, works of pure science aimed at the public have limited sales and no wide cultural impact unless they ride on the back of a successful television series. However, books *based* on scientific research can have huge sales and influence provided they are on topics of interest to the general reader; social topics, ecology, health and fitness are examples of this principle.

There is more scope for science to reach the better-educated members of the public, particularly those with tertiary education. In addition to the more popular works they will read biographies of scientists, topics in the history of science and perhaps certain general works of science. Such readers are of great importance to the scientific community, even though sales to them will never put a book in the fast-seller lists.

In a market-oriented society people read books that have been sold to them as containing benefits they recognize and want. This means that scientists cannot determine in any *simple* way to propagandize their work and ideas. Such an effort would be almost sure to fail except where the product accidentally met a need. If the chosen topic is indeed one of public interest, there are some publishing techniques that are worth thinking about. The use of teamwork in preparing the book, the use of sophisticated graphic design skills, and the use of modern marketing techniques: such ideas should by now be second nature. Another possibility is the use of subscription lists to reach groups of the public with special interests. This method of publishing is standard in certain fields such as law, taxation and investment, and might well be explored in certain areas of science.

Despite all the difficulties, it turns out that the printed book is still an important channel of communication between the scientist and the public, especially the educated public. No other medium allows for the complex and lengthy reasoning so characteristic of scientific argument, and no other medium is so free of legal and practical constraints. In publishing today, the problems are how to identify the consumer groups, how to tailor-make the publications for those groups, and how to sell to them. In future, the emphasis may shift towards ways and means to allow the public access to the kind of information they want.

References

Curwen P 1986 The world book industry. Euromonitor, London

Euromonitor 1985 The book report. Euromonitor, London

Febvre L, Martin J-P 1976 The coming of the book: the impact of printing 1450–1800. NLB, London [translation of L'Apparition du livre, Editions Albin Michel, Paris 1958]

Ivins MW Jr 1969 Prints and visual communication. Da Capo, New York

Kuhn TS 1970 The structure of scientific revolutions, 2nd ed. Chicago University Press, Chicago

DISCUSSION

Laetsch: Some introductory textbooks in biology and psychology sell in much greater numbers than trade books normally do.

Macdonald-Ross: I don't believe that textbooks have much influence until people have become genuinely interested in science.

Lucas: Do juvenile science books such as the Young Observer series sell well?

Macdonald-Ross: I don't know. There are also part-works—magazines that are intended to build up to a complete book—whose influence is probably minimal though still worthwhile.

Miller: The National Geographic Society sells about a million copies of some of its hardback books, covering quite complicated scientific topics.

Macdonald-Ross: Time-Life and the National Geographic Society are both very successful in certain limited areas. Their book production methods are very impressive and we can probably learn something from these organizations.

Miller: The Smithsonian seems to be moving in that direction too.

Falk: The Smithsonian has a very active independent book-publishing arm, Exposition Books, which sells largely by direct mail rather than through bookshops. They specialize in coffee-table books. The *Smithsonian* magazine has been tremendously successful and is a fine example of the public dissemination of scientific material.

Hearn: You suggested that more market research and dialogue with consumers are needed. If there are only 5% or so who are scientifically literate people, will that approach really pay off?

Macdonald-Ross: The public has demonstrated its profound interest in topics that are not purely scientific but have personal, social and other kinds of relevance. Where people are very interested in diet or something like that, we could at least get over the scientific side by being more attentive to people's needs. At the moment the scientific community is not fulfilling that role.

Hearn: I would have thought that magazines are a better vehicle for conveying science than books, in spite of the expenditure figures you gave. Magazines such as *New Scientist*, *Scientific American* and others present a lot of interesting scientific work in fairly easy language.

Macdonald-Ross: I suspect those magazines are immensely important to people who are doing teaching or research in a neighbouring area. The magazines have no impact on the public.

Evered: In the UK *New Scientist* is essentially a magazine for other scientists, as you say, but elsewhere some of the science magazines such as *Sciences et Avenir* and *Scientific American* reach a wider audience.

Tisdell: People in the publishing business have to do market surveys. Publishers sometimes ask writers to write on a particular subject, but that is looking at a static situation. Walter Bodmer earlier seemed to feel that the school curriculum in the UK should be less specialized, with everybody having continual exposure to science and the arts until they go to university. That might expand the readership for some science-type books but it needs a change in society.

Falk: The *Smithsonian* magazine treats a variety of topics in art, history and science. By any criteria it has been profoundly successful but I am not aware of any studies showing that the segment of the audience that is interested in art reads any of the articles on science, or vice versa.

Macdonald-Ross: There is a distinct lack of data on these questions. Funds for research into reading habits have a very low priority. Publishers are a scattered profession, unlike the centralized television industry. Publishers don't in fact do much market research in either the UK or the USA.

Tisdell: Textbook publishers in the USA and Canada do a lot of market research but it would be very expensive to get to the general public in the same way.

Wolpert: C.P. Snow's concept of two cultures may underlie our difficulties. I would guess that in the UK we are dealing with two distinct cultures—those who are scientifically literate and those who are cultured in a quite different way—and that there is hardly any overlap between the two populations.

Friedman: There are cross-over books but there is no way of predicting which books will become a cross-over book. For example, in the USA several books on particle physics have appealed to people's sense of weird philosophy and the books have large sales. The people who read these books tend to say 'I've never enjoyed science or physics before but I read this book because I became interested in Zen Buddhism and actually what was interesting in the book was the physics'. Books that are basically not about science may include a lot of science and attract people to science.

Stewart: As a non-scientist I enjoyed *The Dancing Wu Li Masters*, by Gary Zukav (1979). It opened my eyes to particle physics and the whole world of physics beyond Newtonian physics. I am much in favour of scientists producing that sort of book for a more general audience.

There is also a need for pure science primers for adults. I work in the chemical industry and from time to time I have tried to get to know more about our products. A while ago I tried to find out more about polymer chemistry. In Foyles bookshop the O-level school textbooks didn't seem over-anxious to explain polymer chemistry. In the A-level chemistry section I managed to find the odd chapter but by that time lots of assumptions were being made about the reader's existing state of knowledge. I would like to think that it was possible to talk about polymer chemistry in a few paragraphs to an adult without having to go through the whole story so far. There may be a gradual decline in the self-improvement movement but I think a lot of people want to learn about scientific matters, without the jargon.

Macdonald-Ross: For that kind of study you need to do a computer search and then start by reading the major multi-volume reference works that you cannot bring home with you. Even intelligent and educated people don't necessarily know how to use the library system. Some sort of intermediary between the public and specialists which would tell you how to get this kind of information would be useful but this link is missing at the moment.

Miller: We have been talking about different media for communicating science and there are some natural overlaps that we often miss. For example, several high-selling books were, as you mentioned, derivatives of television shows. Television is a very good vehicle for stimulating curiosity and ideas but it has some inherent limitations to how much it can teach. The same is true for museums. The books people buy in the museum shop are probably the most important form of education. A student of mine found that almost all the people who read science magazines also go to museums and watch science television. That is the smallest group. Almost all museum-goers also watch science television, but not all of the largest group, the science television watchers, go to museums. There is a real hierarchy there which should help us to think about ways of communicating. There is nothing wrong with using television to stimulate reading.

Macdonald-Ross: The lack of variety in television is the problem.

Laetsch: And the amount of time spent watching television leaves very little time for reading. We see this increasingly with our university students—even many of the exceptional students are not readers.

Falk: I have tried to get science books published for the lay public. Trade book publishers are extremely conservative, especially about using marketing research. Trade book publishing has become unprofitable, which has made publishers even more conservative. They are convinced that books on scientific topics are not profitable so they will not publish them. If they won't publish them the probability of such a book becoming popular declines. There is a declining spiral or self-fulfilling prophecy here. Most publishers will publish things they know will sell, such as diet books and spin-offs from television series.

Tan: They look at the actual profit too. If it is too small, there is no incentive for the commercial publisher. To overcome this problem at home, we publish and sell our books at nominal prices.

Thomas: One of the most popular science books published in England in the last 10 years must have been *The Selfish Gene* by Richard Dawkins (1976). There have been many other books like that in which the authors have a particular 'line' to plug, often an element of personal philosophy to put over. Dawkins' book was controversial in certain scientific circles, and there is nothing like controversy to generate interest and get people to read a book. But that is just what often makes scientists reluctant to write books of this kind. They are criticized by their colleagues for writing popular books often enough anyway.

Wolpert: Dawkins was attacked because of particular concepts, not because the book was popular.

Thomas: But the fact that it was controversial helped to make it popular.

Laetsch: As I said earlier, in some universities if you write a popular book you are labelled a journalist.

Friedman: A number of New York publishers are looking for science popularization books. Museum staff often get asked by publishers whether they have any manuscripts. The publishers want books that are readable and not too long, with sexy-sounding titles such as 'Adrift in the cosmic river'.

Art museums tend to have expensive books to go with their blockbuster exhibits. The production of these books is often heavily subsidized out of the grants for the exhibits. Except for the illustrations these books are often inaccessible. They are nevertheless bought by a lot of people. Are they read, and would they work with science exhibits? I have seen no research on the readership of these art books but they are printed and sold in tens of thousands.

Serrell: What are the sales of books that back up your science exhibits, Dr Miles?

Miles: The books of our most popular exhibition, Human Biology, have sold about 100000 copies around the world and have been translated into three or four languages. Our first five new-style exhibitions were accompanied by similar books as a major educational element in the whole exercise, and these have also been translated into various languages. They were co-published with a commercial publisher. We have had to stop producing such books with recent exhibitions because they take up too much of the design team's time. The commercial publisher nevertheless would be delighted if we started again. One reason is prestige. Commercial publishers like their name attached to that of the Museum, and of course they make money.

Gregory: Some publishers publish books they know will lose money.

Macdonald-Ross: Such books don't reach the public.

Gregory: There are also companies like Dover that produce reprints. I think a lot of people buy these.

Macdonald-Ross: Facsimile reprints are now quite cheap to produce. That doesn't tackle the problem of communicating new science but it does allow interested students to read some of the classics of science history.

Laetsch: A lot of attention has been devoted in the USA recently to the number of people who are functionally illiterate. I wonder if this is a constant in contemporary literate societies? General literacy rather than scientific literacy may be the real problem. If we reach a certain standard in general literacy, scientific literacy might take care of itself. If one in five people are functionally illiterate in literate societies our discussion is revolving around different flavours of cake rather than basic nutrition. To what extent can science be used as a vehicle to increase the functional literacy of people?

Macdonald-Ross: The research into functional literacy has been much more impressive than the research into general readership (Sticht 1975). One way of tackling the literacy problem is to improve the basic standards of design in communications to the public. For people on the borderline of literacy, badly designed documents are a great barrier to understanding. It is much easier to change the documents than to change the people.

Laetsch: In the USA it is predicted that one in 10 individuals are in the spectrum of dyslexia, which could explain a lot.

References

Dawkins R 1976 The selfish gene. Oxford University Press
Sticht TG 1975 Reading for working: a functional literacy anthology. HumRRO, Arlington, Virginia
Zukav G 1979 The dancing Wu Li masters: an overview of the new physics. William Morrow, New York and Fontana/Collins, London

The influence of pseudoscience, parascience and science fiction

ALAN J. FRIEDMAN

New York Hall of Science, 47–01 111th Street, Flushing Meadows-Corona Park, NY 11368, USA

Abstract. Every day some science and some material resembling science reach the public through those mass media which present pseudoscience, parascience and science fiction. The daily astrology column run by most newspapers in the United States is a familiar example. Studies in several countries offer evidence that these presentations have a substantial impact on the public's images of science and scientists. Science educators often see science fiction in a positive light, as generating interest in science and technology. Parascience and pseudoscience, in contrast, are usually viewed as a hindrance to science education. The most cited instance is the attempt by advocates of 'creation science' to influence the teaching of evolution.

Educators in schools and museums disagree on what to do, if anything, about parascience and pseudoscience. Some planetarium educators believe they should ignore astrology to avoid dignifying it; others believe they should confront astrology directly, contrasting its goals and methods with those of astronomy. Two current arguments question (1) the wisdom of spending any of the limited resources of science education on countering pseudoscience, and (2) the effectiveness of rationally criticizing non-rational beliefs.

1987 Communicating science to the public. Wiley, Chichester (Ciba Foundation Conference) p 190–204

Alternative channels of public communication about science

Mass media information about science, and about topics resembling science, reaches the public each day. Some of this information is prepared by professional science communicators such as the teachers, museum staff, science journalists or research scientists who have come to this Ciba Foundation Conference. But much of this information comes from people who are not members of this community of professional science communicators. These are the people who present pseudoscience, parascience and science fiction.

I wish to review the influence of these three categories of public information on the overall communication of science to the public. These do not yet exhaust all the channels of science communication, of course. There are more casual, less explicit presentations: advertisements which purport to show valid experiments proving the superiority of one brand of paper towel over another or the efficacy of an over-the-counter drug. Much scientific or technical information is also dispensed in the form of articles, TV programmes and books on weapons research, high technology business, consumer product reviews, and household repair hints. I have selected pseudoscience, parascience and science fiction for discussion because these three channels explicitly claim to be science or to have serious connections with science.

Pseudoscience, which includes astrology, creationism and Erich von Daniken-variety archaeology, claims to use some of the tools and concepts of science, but scientists in the traditional academic and scholarly world find that the supporters of these pseudosciences do not follow the accepted rules of evidence and other critical standards. Parascience denotes fields on the fringes of established science which have attracted pseudoscientific interest— theories of the relations between quantum physics and consciousness, investigations of unidentified flying objects, parapsychology. Science fiction is a literary form and does not claim to be a science, but science fiction nevertheless mixes established science with speculation, in a fashion that resembles some of the pseudosciences and parasciences.

The public is exposed to substantial amounts of pseudoscience and parascience in books (Dutch 1986) and newspapers (Kurtz & Fraknoi [1985] estimate that 1200 US newspapers run astrology columns). Newspapers that emphasize pseudoscience have large circulations: the weekly *National Enquirer* has a paid circulation of 4 552 047. A typical issue has stories such as 'UFOs Are Routine Sight Above Indian Reservation . . . One Witness Has Seen Hundreds' and full-page advertisements offering to 'Improve Your Life Now . . . [with a] scientifically designed MAGNETIC STIMULATION BRACELET' (23 September 1986). By contrast, *Discover* is the largest circulation popular science monthly in the USA, at 923 130 copies, and *Science News*, a reputable weekly, has a circulation of 155 832 (data for 1985). Some nationally broadcast television programmes, such as the long-running 'In Search of . . .' series present uncritical accounts of pseudoscience.

Even conservative, establishment institutions may take pseudoscience seriously: an article in the business section of *The New York Times* covered business leaders who consult psychics and astrologers: 'Mr. Attride and other serious astrologers go way beyond the ubiquitous newspaper horoscopes . . . most will customize a horoscope for clients . . . "An astrologer will compare the birth chart with current conditions in the solar system and make predictions, business and otherwise," said Mr. Attride' (Kaufman 1985).

The fashion magazine *Harper's Bazaar* runs a regular astrology column and recently printed an article announcing that astrology is becoming a respectable aid to psychotherapy: ' "In a way, getting an astrological reading is like getting an X-ray", says Dr. Barbara DeBetz, assistant professor of psychiatry at Columbia University and a Manhattan psychiatrist. "It provides a baseline picture of your personality, which alone can be therapeutic for some" ' (Merlin 1985).

Does this material have any substantial impact on the public's understanding of science? There is no accepted position among science communicators on the amount of seriousness of the influence of these alternative channels. From discussions with several dozen individuals from my own profession (museums and planetariums), however, I find some common opinions and concerns on each of these alternative sources of public information about science and technology.

The influence of science fiction

Science fiction often contains accurate science, presented with pedagogical clarity. Some scientists and engineers attribute to science fiction their first childhood interest in science. Science fiction books are almost always careful to distinguish between well-accepted science, contemporary theories and imaginative projections. Science fiction is used in some schools and universities as a means of introducing students to science itself. For example, Poul Anderson's *Tau Zero* (1970) and Joe Haldeman's *The Forever War* (1974) are accessible presentations of some of the spectacular implications of relativity theory (see also Williamson 1980, Freedman & Little 1980).

The images of science and of scientists presented in science fiction cause some concerns, however. Scientists are often presented as super-geniuses, idiosyncratic, out-of-touch with normal people, and apt to cause harm to society (if inadvertently). These images of scientists in science fiction are similar to the image of scientists held by many students. A number of studies of the image of scientists accompanied the Sputnik-era rush to recruit students into science. A national survey sampled essays on scientists by 35 000 students. A composite of common negative statements about scientists included: 'He neglects his family—pays no attention to his wife, never plays with his children. He has no social life, no other intellectual interests . . . he has secrets he cannot share . . . He is always running off to his laboratory. He may force his children to become scientists also' (Mead & Metraux, 1957; compare this with a study of the image of the scientist in science fiction by Hirsch [1958]. For a more recent study, see National Assessment of Educational Progress [1979]).

While there are certainly exceptions to these stereotyped presentations of scientists in science fiction (for example, Benford 1980, Preuss 1985), it is arguable that the stereotypical picture of the scientist in most science fiction is damaging to the public understanding of how science interacts with society (Friedman & Donley 1985).

Influence of pseudoscience and parascience

Science educators generally accept that a significant percentage of the public believes many of the claims of various pseudosciences. While degrees of belief are difficult to quantify and the wide range of topics described as pseudoscience or parascience are poorly defined, many surveys indicate that belief in the validity of pseudosciences is widespread (Bainbridge 1978, Adelman & Adelman 1984). A recent Gallup poll showed that 55% of American teenagers believe that astrology works (Frazier 1985). Even a successful candidate for the office of President of the United States has professed some belief in astrology: 'Reagan says he follows the daily zodiacal advice for his sign in the horoscope column of Carrol Righter . . . "I believe you'll find", he said, "that 80 percent of the people in New York's Hall of Fame are Aquarians" ' (Dunn 1980).

A survey in France indicated that many adults regarded astronomy and astrology as two branches of the same science, one concerned with physical phenomena alone, the other with the impact of those phenomena on people (M. Thiesse, unpublished seminar, Parc de La Villette, 1982). Among the reasons people gave for believing that astrology was a science were that astrologers use computers, just as astronomers do.

A belief in 'creation science', as well as a disbelief in evolution, are often cited by science educators in discussing the recent wave of laws proposed or enacted in the United States—Arkansas, Louisiana, Minnesota, Tennessee—mandating the teaching of creationism or hampering the teaching of evolution in schools, or both (Zetterberg 1983). Such laws have had a major chilling effect on the teaching of evolution in many parts of the United States. That scientists regard this challenge to science communication as serious is indicated by a recent legal brief filed at the US Supreme Court by 72 Nobel Laureates and 24 scientific organizations urging the court to strike down the Louisiana law (Taylor 1986).

Educators discuss other, less dramatic instances of pseudoscience interfering with science communication. Some planetarium educators note that both school and public audiences expect astrology in the planetarium, and are disappointed to find that only astronomy is offered. Overcoming false expectations occupies much of the time that could be spent in presenting science. Claims of breakthroughs and revolutionary discoveries by pseudoscientists distract students, several educators say, from the less spectacular and more subtle achievements of science itself.

Some educators report being put on the defensive by attacks on science and scientists from proponents of pseudoscience. Creationists and other pseudoscientists point out inadequacies and continuing revisions in evolution and other scientific theories, as if these demonstrated the failure of conventional science. Time discussing these well-publicized challenges to science must be at the expense of time that could be spent discussing the advantages of the self-critical, self-correcting nature of science and its real achievements. Nevertheless, failure to deal with these challenges, some educators believe, may lead students to think that science is ignoring just criticisms, and is perhaps ignoring valid sciences which happen to disagree with establishment science.

Educators are also concerned with the possibility of a general trend towards non-rational or irrational beliefs, inspired by the constant appeals of pseudoscience to reject establishment science (see introductions and articles in Schultz 1986; and see Abell & Singer 1981, Kurtz 1985).

What is being done by establishment science communicators

With some exceptions, the established channels of science communication to the public ignore pseudoscience. A group of 186 scientists published a statement expressing concern over public acceptance of astrology (Bok et al 1975), and scientists have gone on record opposing legislation supporting creationism (Taylor 1986). But planetariums do not often mention astrology except as a historical footnote. Few science museum exhibits treat pseudoscience in any fashion. Criticism of pseudoscience is absent from most school curricula. Science fiction is used as an aid to teaching, but little effort is made in most classrooms to overcome the common stereotypes of scientists in science fiction.

Science communicators give several reasons for ignoring pseudoscience, parascience, and the stereotypes of science fiction:

For whatever reason, people *want* to believe in pseudoscience; therefore it is essentially a religion, and it is not the business of science to criticize religious beliefs.

Pseudoscience is irrational; rational argument cannot counter it successfully.

There might be something in the mass of pseudoscience that is correct but at present unknown to science. We would be foolish to attack pseudoscience if even a tiny part of it proves out.

Unlike creationism, astrology does not attack science itself. Live and let live.

We have enough to do communicating real science; there is no time to attack false science or correct inaccurate perceptions about the personalities of scientists.

We would dignify pseudoscience by mentioning it at all.

The minority of science communicators who take pseudoscience seriously use primarily one of two approaches. The first is a careful, historical presentation, which shows how science diverged from other forms of knowledge by formulating a systematic experimental method, with rules of evidence. Astrology is treated as a case of arrested development, an interesting hypothesis before Kepler and Newton, but failing to progress since then. The other approach is a direct treatment of pseudoscience through the application of rigorous tests to its claims. Dozens of careful studies of astrology and other pseudosciences are themselves educationally useful case studies of scientific methods (see issues of the *Skeptical Inquirer* or collections like that of Abell & Singer 1981). This approach also demonstrates that science must and can be open to new ideas, even if they seem unlikely.

The argument for doing more

Some science communicators take the popularity of pseudoscience as a symptom of a more serious problem, an acceptance of soft thinking in general and a lack of understanding of science specifically. I find these concerns compelling. The reader of the *National Enquirer* who accepts claims of the UFO story uncritically is also likely to accept the claims for the 'Miracle Stimulation Effect' of the magnetic bracelet as a cure for health problems. A public that has little interest or faith in careful argument is potentially fatal to a democracy in an age when science and technology are increasingly present.

Much contemporary presentation of science—in classrooms, museums or other media—treats science as entirely a growing collection of facts, handed down by unseen geniuses. This treatment allows for no acceptance of uncertainty, no continuing role for intuition or inspiration, no appreciation for scientists as human beings with feelings, hopes and carefully developed (if imperfect and sometimes painful) techniques for self-correction. This presentation of science makes it easy for pseudoscientists to present themselves as a more open, thoughtful, fully human breed of investigator.

I would like to see more of us take up the challenge of pseudoscience and mis-imaged science. We can present science as a human process beginning with an open curiosity and intuition and producing a cumulative body of visions and data. The careful and gleeful work of debunking fraud and fuzzy pseudoscience is one part of this. The faith in reason and the joy of creation that attracted us to science in the first place are other essential elements we can try to communicate to increase the public's understanding of science.

Acknowledgements

I would like to thank the dozens of my colleagues at the 1986 International Planetarium Society meeting who offered me their thoughts and concerns on pseudoscience. I am also grateful to Andrew Fraknoi, Linda Hechtman, Jonathan Herman and Priscilla Watson for providing me with many useful references.

References

Abell G, Singer B 1981 Science and the paranormal. Scribners, New York
Adelman AS, Adelman SJ 1984 Pseudoscientific beliefs of sixth-grade students in the Charleston, S.C., area. Skeptical Inquirer 9:71–74
Anderson P 1970 Tau zero. Doubleday, New York

Bainbridge WS 1978 Chariots of the gullible. Skeptical Inquirer 3:33–48
Benford G 1980 Timescape. Simon & Schuster, New York
Bok BJ, Jerome LE, Kurtz P 1975 Objections to astrology. Humanist 35:4–6 (September/October)
Dunn AF 1980 I was always the last one chosen (interview with Ronald Reagan). Washington Post E5 (11 July)
Dutch S 1986 Four decades of fringe literature. Skeptical Inquirer 10:342–351
Frazier K 1985 Gallup youth poll finds high belief in ESP, astrology. Skeptical Inquirer 9:113–115
Freedman RA, Little WA 1980 Physics 13: teaching modern physics through science fiction. American Journal of Physics 48:548–551
Friedman AJ, Donley CC 1985 Einstein as myth and muse. Cambridge University Press, New York, Chapter 6
Haldeman J 1974 The forever war. St Martin's Press, New York
Hirsch W 1958 Image of the scientist in science fiction: a content analysis. American Journal of Sociology 63:506–512
Kaufman J 1985 What's new in parapsychology. New York Times F19 (3 November)
Kurtz P (ed) 1985 A skeptic's handbook of parapsychology. Prometheus Books, Buffalo, NY
Kurtz P, Fraknoi A 1985 Scientific tests of astrology do not support its claims. Skeptical Inquirer 9:210–211
Mead M, Metraux R 1957 Image of the scientist among high-school students. Science 126:384–390
Merlin K 1985 Astrology as therapy: from couch to constellations. Harpers Bazaar, p. 80–106 (October)
National Assessment of Educational Progress 1979 Report 08–5–02 (Attitudes towards science). Denver, CO
Preuss P 1985 Human error. Doherty, New York
Schultz T (ed) 1986 Special issue: the fringes of reason. Whole Earth Review, No. 52
Taylor S 1986 72 Nobelists urge court to void creationism law. New York Times A17 (19 August)
Williamson J (ed) 1980 Teaching science fiction. Owlswick Press, Philadelphia
Zetterberg P 1983 Evolution versus creationism: the public education controversy. Oryx Press, Phoenix, AZ

DISCUSSION

Gregory: Kepler was an astrologer and Newton believed in alchemy. Why do we now regard these subjects as irrational? We think of the structure and function of the brain as conveying meaning, yet we deny this for the stars. But deciding what is occult and what is not, is not a simple matter. We cannot just say that the paranormal phenomena we don't understand are not science. There is a lot that we don't understand in what we agree is science. In physics, for example, we don't really understand magnetism. We seem to be saying that these areas will never be part of science—we are legislating about the *future* of science in order to classify the paranormal. We really have no sound intellectual premises for classifying pseudoscience.

Wolpert: I disagree. The real difficulty is that there is no evidence for the paranormal. It can't be that difficult to demonstrate levitation but no one ever does. I think people are keen on the paranormal because it makes them scientists. In our kind of science it takes years to establish the most trivial facts. In the paranormal you can make the most amazing discoveries in seconds, so everyone can be a scientist instantly. I think a lot of these people feel they are excluded from the hard world of real science and that this is one of their driving motives.

Gregory: I agree, but that is not what I was saying!

Whitley: The boundaries of parapsychology have shifted over the past 40 years in different societies. The role of psychologists and statisticians in evaluating Rhine's experiments in the 1930s was mutually contradictory. I am amazed that at a meeting like this you can produce labels of this kind as if you were describing something that is important. What is the importance of the label? What is the theoretical problem you are trying to address? The whole meeting seems to have refused to address the questions of why we should communicate science, what aspects of what science we should communicate, and to whom. What is the focus of concern that is engaging us?

Wolpert: I don't think the paranormal has changed at all over the last 40 years. One of the characteristics of what Langmuir called pathological science is that it doesn't change. It matters to me because I have to work so hard in my science and the paranormal believers don't have to work. Secondly it really is anti-science and it undermines serious thinking about the nature of the world we live in. Science is the best way we have of trying to understand how the world functions. The pseudosciences undermine that enterprise.

Whitley: You are assuming that everybody knows what these sciences are and that they are remarkably stable over time and over discourses. That is precisely what I am querying.

Friedman: There are boundary areas and the boundaries shift dramatically in time. Pseudoscientists often cite the failure of natural scientists to believe for a long time that meteorites really come from outer space. My definitions of current pseudoscience are based on those in the *Skeptical Inquirer*.

Whitley: There has been a lot of debate about the evidence used in those articles.

Friedman: The notion that characteristics of individuals and their personality types can be determined by the sun sign at the moment of their birth is the most popular form of astrology. *None* of the excellent studies that looked for correlations between sun sign and numerous possible personality characteristics or events found any statistically significant correlations. I therefore call sun-sign astrology a pseudoscience. Are you arguing that there is *some* evidence for its validity?

Whitley: You are assuming that there is something essential about science that we all know and understand and that this has been universally true. My point is that you must put that into context and say under what circumstances do which activities become regarded as scientific. And why is it now more important to be regarded as scientific than it was in the 18th century? What do those terms mean?

Suggesting that if something is statistically significant it is scientific is a very dubious claim. Much research in social science makes a great claim of statistical significance but this is not what 'real scientists' such as Lewis Wolpert might regard as scientific. Social scientists have a different concept of what is scientific. Statistical significance on its own is meaningless.

Friedman: Again, are you arguing that there is some evidence, statistical or otherwise, in favour of sun-sign astrology?

Whitley: The idea that certain rules of evidence decide whether knowledge is claimed as scientific or non-scientific is itself a claim which has to be justified by reference to a broader framework of discourse. You can't get away with asserting that if it is significant statistically it is or is not scientific.

Friedman: That is not what I said. I said that there is no evidence showing a significant correlation between sun-sign astrology and personality. Is there other evidence to show a correlation?

Whitley: I am not an expert on sun-sign astrology.

Miller: This objection illustrates part of the problem. I think you are arguing that if every idea is of equal merit, one idea should not be discounted. If we think of science as a method by which we can know certain things and if at any given time there is some consensus on a certain body of knowledge, then although that consensus may change over a period of decades or centuries, we suggest that it doesn't change over hours and days in quite the way that sun-sign astrology and other things may change. You seem to be arguing that there is nothing that can be right or wrong.

Whitley: I am not arguing that at all. As soon as I say that knowledge claims are not automatically true just because they come from people who are accredited scientists you seem to think I mean that all knowledge claims are equally valid. But I am pointing out that the use of statistics to make particular knowledge claims is dependent on the framework within which those statistics were drawn up in the first place, the theory from which they are meant to be derived, and the purposes for which they are being used. I am simply saying that statistical significance on its own is totally meaningless.

Tisdell: You are saying you are an agnostic, which is fair enough. At present there is no evidence for sun-sign astrology but if someone produces evidence for it in the future I might change my mind and see some value in this astrology. The important point is that it is up to scientists to speak out and say that there is no scientific evidence, no statistical evidence, to support this claim. I would be inclined to accept the statistical test and most rational people ought to be inclined to do that. Otherwise there is the danger that we may say everything is equally acceptable, which is complete nonsense. On the other hand I am prepared to change my mind on some things if someone produces adequate evidence.

Laetsch: Is some of the problem due to our ability to hold several belief systems simultaneously and slip back and forth between them? People can practise and accept all the rational rules, as we define them, of science but at the same time believe or practise the contrary.

Friedman: I don't think it is a matter of holding conflicting beliefs simultaneously. People who write textbooks either do or do not believe that it is valuable to present Darwinian evolution but some of them are specifically prevented from writing about it.

Laetsch: That may be part of the problem. The books present natural selection as an interpretation of evolution without defining it as a belief system that may or may not be better than another belief system.

Lucas: What cues do readers or visitors to exhibitions or listeners to radio programmes use to decide what is parascience, what is pseudoscience and what is science? A science programme on Australian radio fairly regularly has a spoof item which the people in the know quickly recognize as a spoof. I suspect that the people who are not in the know may not recognize which items are jokes. How do you distinguish experiments on pyramidal sharpening of razor blades and on the search for gravity waves when you hear the descriptions without knowing about either of those things? What cues can programme-makers or journalists build in to authenticate the information?

Deehan: We are very careful . . .

Rhodes: One could argue that the occasional spoof item ought to sharpen the minds of the listeners or viewers and train them to assess whatever is presented.

Lucas: I am not convinced that people always recognize it as a spoof item. We don't know enough about our audience to know what proportion will detect the spoofs.

Deehan: We have had two items in the last 10 years that might be regarded as spoof items. The most recent was the coin-operated defibrillator in a hospital. We stated at the end that it was a joke. A strong lesson in broadcasting is that unless you are very careful people will hear or see only what they want to hear or see.

As for authentication, we apply the same standards of evidence as we use for mainstream science to anything else that comes along. If I am not convinced myself, I ask scientists about the item. At some point we have to make a judgement about whether something is reasonable.

Hearn: The distinction between science and pseudoscience might be something we can approach in this discussion but the nature of evidence is not approachable by the general public looking at a science programme. If Uri Geller bends a spoon or draws a house which happens to be the same as a drawing Terry Wogan has in his pocket, that, for the general public, is first-class evidence in favour of parapsychology. Getting around that is a real problem.

At the London Zoo a couple of years ago, on April Fool's Day, a man put on a strange primate costume and it was reported on television that this was a new species found in the wild. Many people came specifically to see this animal and were annoyed when they learnt it didn't exist. People still phone to ask whether the zoo still has the new animal they saw on television.

Macdonald-Ross: If we really want to get over the *method* of science we have to step back and find a way of communicating with people that enables them to take part. Instead of telling them 'This is not science' we can say 'This is how you might think about it, now what do you think?'

The mass media gave almost no publicity to the professional magicians who knew that Uri Geller did his tricks by traditional prestidigitation. On this particular occasion the media did a bad job for us.

Rhodes: The problem with Geller is to get him to participate in anything serious.

Hearn: The question Mr Macdonald-Ross is asking is perhaps the same as mine: how do we get public participation in interpretation and in learning the method of science?

Macdonald-Ross: That is much more at the heart of what we would like the public to know about science than questions of the latest discoveries.

Tisdell: The figures that Alan Friedman gave for newspapers carrying astrology columns seem frightening. On the other hand many readers look on them as fiction; they know they are not serious. You have to allow for that. The dangerous people are those who read your palm and you really feel they are predicting your future or something like that.

Friedman: There are establishments which ask $250 to produce a computerized horoscope. When people are ready to hand over that kind of money I would credit it more than you do.

Tisdell: Yes, but more research is needed to find out whether it is a true belief system.

Lucas: In the MORI poll that we did, 75% of the British population accepted the notion that astrology was scientific. They were not confusing it with astronomy. But our quesion asked 'was it scientific?', not 'was it a science?'

Gregory: What science says of the world is totally counterintuitive. We are selling something which is rejected by people on commonsense grounds. The difficulty of getting science across is colossal. When we get to cognitive psychology we can't even justify it within science. How decisions are made in a bunch of brain cells is very mysterious and it looks occult to any intelligent neurophysiologist. We don't have a sensible physical model of information handling, either in a computer or in the brain.

Miller: An issue that several people have been raising is how to get the public to understand the rules of evidence for the procedures we have been talking about. It is extremely difficult and expensive to get adults to reverse their thinking patterns. An important rule of school science is that people should be immersed in the notion of evidence as a scientific approach, so that they will indeed ask some of those questions. School science should not consist of memorizing the names of hundreds of minerals or other things. It should emphasize evidence and what makes science different from other kinds of belief systems. In most societies, schools are the place where it is most cost-effective to deal with that question. Everything else is expensive and difficult and has a low probability of success.

Laetsch: But how long would high-school principals last in many places in the United States if they were accused of teaching science in order to expose religion? There would be a terrific community problem, because our schools are run by local school boards.

Miller: You can certainly teach scientific method. If you attack religion directly that is a matter for public discussion.

Hearn: A recurrent theme in this meeting has been that we should push everything back into schools but I very much doubt whether we can even teach students to discriminate between pseudoscience and real science. In a way we are expanding the problem. We need to get to the adult general public too. The school side is extremely important but it has to go well beyond that. If someone asks me what I did best at school I would say rugby and boxing.

Bell: There is a growing movement towards allowing students to get their own questions answered. It is found best in primary science but it is just as good in secondary science. It is only when people test things for themselves that they change their ideas, not when they follow instructions that someone else has given them.

Friedman: I would like to see science education leaving at least a modicum of scepticism among the general public, so that they would query sun-sign astrology and other areas for which there is no evidence or for which there may be clear evidence of fraud. It is not impossible to instil scepticism in the general public. But clearly the format of science education we have used—learning a collection of facts produced by unseen geniuses—seems to discourage that kind of scepticism.

Wolpert: Psychology has shown that we tend to make judgements on limited evidence. One or two instances in context will persuade us to all sorts of false conclusions and our brains do not work easily on statistical evidence. We are being wildly over-optimistic in trying to overcome a basic human quality. Most of us have superstitions and non-rational attitudes.

Duensing: A fundamental process of human perception is to create patterns of meaning in nature. It is something we continuously do. But if you perceive a pattern that doesn't match what someone else says is true, or if you make an incorrect connection between things, how do you prove yourself wrong? It is very hard to do. Even though it is hard to verify what is accurate, it is important to encourage people to notice patterns, to make connections for themselves. Frank Oppenheimer used to say, 'misconceptions are better than no conceptions at all'.

Falk: There is some evidence that humans and other organisms try to maximize their control of their fate. Adaptively you want to avoid making a big mistake. If a safe is going to fall on your head while you are walking down the street that is a calamity to be avoided at all costs. If your horoscope says, for example, that if you go outside today you will get hit by a safe, then even though you may know that the probability is against it, you might still stay inside rather than risk going out.

Friedman: Would you spend money on a magnetic bracelet instead of visiting your physician?

Falk: Yes, if my belief system seemed to reinforce the idea that that would increase my survival chances. I might do both! Copper bracelets may not do anything, but on the other hand they also might help.

Miller: In schools, we should try to teach science in terms of probability. I can teach statistics to children of 12–14 years of age. At that age they like probability as a construct. They like flipping coins, for example, and you can begin to deal with probability. I think that kind of experience changes one's perspective on evidence. Most people can do these things if someone shows them how.

Falk: The success of science as a discipline ties into the notion of fate control. That is why western society by and large has been successful. Through this process that we all buy into, our ability to control our environment seems to be enhanced.

Rhodes: We should be looking at the areas where we can do things effectively. There is a sharp end and a blunt end and the sharp end (radio and television) can easily be turned off. In zoos or museums you have a couple of hours to produce an effect. In schools you have students there for a number of years and they can't walk out of a door. How do you deal with this marvellous captive audience? In my school days the system failed to teach me to speak French, although I studied it for six years. I suspect that happens in science classrooms too, but we can't simply ignore the 80% of students who are apparently going to become the scientific illiterates. There is an interest there and a demand for information. The schools are the area that we should be discussing. How you label an exhibit in a museum is irrelevant to the main issue of the public understanding of science.

Laetsch: There is some disagreement on that point. Food faddism of various kinds illustrates our problems. Food faddism is based on bits and pieces of what was considered science by the health profession at different times. Much of the public debate has consisted of nutritionists and so forth attacking each other in the press over what one should or should not eat to avoid cancer. The public puts its own mix into this and says it is all nonsense. Many scientists pay little attention to this debate because we consider that one belief system is superseding another. There are real problems about how scientists themselves debate things.

Miller: In one of our surveys, we asked people what they thought about astrology and whether they changed their behaviour on a given day because of their horoscope that day. Five per cent of the adult population (representing 8.5 million people in the USA) said that they changed their behaviour. We also asked them later in the same interview whether they had followed issues about the carcinogenic character of certain foods and if they ever changed their eating or buying habits. Fifty per cent said they had. So at least in the United States a lot more people follow what they believe to be science than follow astrology.

Stewart: As the 'resident non-scientist' at this meeting I still feel like an outsider after two and a half days. I think that is simply because I am a non-scientist, and that does concern me. As a group you have been very closed and when it comes to communication that is worrying. Communication, as David Evered said in his invitation letter, is a two-way process. It is absolutely vital in communication to look at things from the recipient's point of view. I am not sure that enough of that has happened in this group over the last two and a half days—looking at science communication from the lay standpoint.

Another thing is that classification is vital in your disciplines, yet for most of the general public the difference between technology, science, engineering, the pseudosciences and so on is about as important as theologians discussing the difference between consubstantiation and transubstantiation. What is important is credibility. When scientists speak to the outside world they need to be believed. When a major disaster occurs, natural or otherwise, a lot of people immediately become interested in the agent of destruction. After Bhopal it was methyl isocyanate, after Seveso it was dioxins, and after Chernobyl it was radiation. Slightly less in the public domain is the debate on whether formaldehyde is a human carcinogen.

Science has let us down in the outside world in not being able to get its act together to present a consistent scientific view on these subjects. With the scientific method one would think it would be quite easy to come to irrefutable logical conclusions about a given set of data. In practice this doesn't happen and therefore people don't know whom to believe. They don't believe scientists because they are arguing among themselves all the time. If you are going to get communication right, credibility is the most important thing.

I don't think that during this meeting we have made out a special case for the communication of science as distinct from any other part of human knowledge. What I have been saying could have been said about any other subject. Scientists ought not to put themselves on intellectual pedestals: their problems are not unique. If we substituted 'art' for 'science' in the programme for this conference, all the other words could stay the same.

In spite of all their jargon, scientists love to call whatever they have invented by some endearing acronym which makes it sound totally innocent. An acronym might help this conference and the one I suggest is LASBOC: let all scientists be open and credible.

Wolpert: What you say about art is fine but you would never say of art that you want it to be credible, nor would you suggest that all the artists should get together because artistic matters will lead you to a credible conclusion. You would never say those things of anything other than science. What worries me is that your image of science makes you believe there is a method that, given the data, leads to an unequivocal result. We certainly need to change the public image of science, because that is not what it is like at all.

Chairman's summary

W.M. LAETSCH

University of California, Berkeley, CA 94720, USA

1987 Communicating science to the public. Wiley, Chichester (Ciba Foundation Conference) p 205–207

A lot of communication is going on. Roger Miles says that 25% of the population of the UK has visited the Natural History Museum. That is, a large proportion of the population has hit a particular educational target at some point, so maybe labels are important in terms of the national life. Whatever people's motives or whatever the outcomes of their experiences, there is an insatiable interest in things related to science. One example is that when Berkeley's Lawrence Hall of Science developed a science exhibit some years ago that travelled to large shopping malls in different parts of the country it competed very effectively with all the other attractions of the malls. A first principle in communicating with the enormous variety of people who are interested in different aspects of science is redundancy. We need lots of different ways to communicate with lots of different people.

I liked Ian Stewart's medical model of the recipient—the idea that we treat the consumer as a patient in matters of science and say that the doctor knows best. In many cases that doesn't work. Science communication has generally been one-way. Chan Screven's comment that people surprise you supports this. It is dangerous to make too many definitions and too many *a priori* statements about what people are or are not capable of. A whole mix of people want information about science. We talked about science literacy being important for the function of the democracy. On the other hand public science communication is a very democratic process. We should not be too limited in whether and why and how we communicate.

Lewis Wolpert said that different kinds of scientists do things in different ways. This means we must be careful about how we define scientific literacy. We must also remember that scientific literacy changes in both space and time. An individual may be literate in one area and not so literate in another, and this can change over time. We tend to become confused, particularly in formal education, because of what I call the liver pâté view of education: if you stuff grain into a goose for a certain time you can build up a certain size of liver and collect it at a particular time. Formal education assumes that if you learn something you will be ready for the next stage at a particular time. But learning does not happen in that fashion. A person may be scientifically literate in a particular way at one time, then forget everything and have to have it triggered again or re-learn it later. We must be cautious in defining scientific literacy and we must not develop the notion of insiders and outsiders.

Jon Miller and others asked frequently 'Is it science?'. We could argue for weeks on the proper definition of science. One of the problems in public science education or anything connected with the mass media is that the medium becomes the message. The equilibrium tends to shift in that direction because it is so hard to get people's attention. If you don't get their attention they turn the dial off and all your good intentions or wonderful productions are wasted. But getting their attention is not all you should do. You then have to provide good science.

We know a lot about how to communicate science, so why don't we use effective communication more generally, particularly in formal education? The informal education that we mostly talked about is consumer-oriented, while formal education is producer-oriented. What clues can the former give us about communicating more effectively in the latter? Formal education is where we have to do things related to a deep or even a reasonable understanding of science. As Lewis Wolpert said, science is not easy or attractive and it takes intensive effort over a period of time.

Another difficulty in applying what we learn about communication in informal science to formal science is that certification drives education in the direction of being producer-oriented. I am not sure we can ever get away from that. Along with that is the hairshirt view of education which views education to a certain extent as a monastic activity. It is hard, difficult, unpleasant and associated with a very formal authoritarian way of doing things. This is part of our historical baggage. If something is interesting or fun there is obviously something wrong with it. In the United States if we apply the lessons we learn from informal education to teaching in schools, the parents after a while will not let it continue. It is not considered rigorous, so it can't lead to certification or to university slots or jobs.

John Hearn mentioned that much of our professional science communication at meetings is now done by what people in museums call a panel show, by putting a modified textbook on the wall. This is historically the way museums have tried to interpret things. How to present information effectively in these panels or posters is not a completely trivial matter. If you want to communicate information it is as important to get the attention of professional scientists as it is to get the attention of non-scientists. In scientific meetings, a common feature is boredom. It is difficult to listen to more than a couple of papers a day. We also ignore the social context. Panel sessions often act as an attractant for people who then interact socially.

Understanding the social context helps to define the ways of measuring and looking at things. Up to about 10 years ago, informal science education was evaluated on the basis of psychometric paradigms. This was very inappropriate, because we were not measuring anything in the context of what people did. When I espoused evaluation methods based on the methodology of ethnologists and ethologists, I was pretty well savaged, but these methods are now taken for granted. The social context also defines the cultural context of how we present the science and the way in which the consumer uses the institution. In other words, we can say what we want museums or television programmes or publishing houses to do, but the public really defines what those institutions are.

The audience defines what and how we communicate. If you want to communicate something effectively, you have to know whom you are trying to communicate with, what their interests are, and the most effective way of getting to them. Formal education pays far too little attention to this. Learning is looked on as an isolated activity. Even though people are now looking at the cultural context of learning in schools, this hasn't got very far.

Richard Whitley asked whether we had made an argument for scientific literacy among the general public. I suspect that very large numbers of the public are interested in science and that this is an imperative that we will fall into. The imperative for scientific literacy doesn't discount the imperative for literacy in other areas at all. We haven't talked much about how effective communication or communication in general influences what we do in science. We have mostly discussed things that could be one way. The Royal Society report is very much a one-way model but communication can influence what we do in science. The general public notion that what is interesting in science relates only to computers has had an enormous impact on the numbers of students who want to study computers. There is also a new wave of students who say they want to study genetic engineering. In newspapers and on television genetic engineering is presented as something very important.

Communication is very causative. It is causative in politics, in economics, in entertainment and so on, so why shouldn't it be causative in science? Communication will increasingly determine what science is, even in the scientific community, and what scientists themselves do. That is, communication is not from science out to the consumer but two-way. What goes out is going to feed back and influence what we do, whether it is trying to get more money from the government or in all kinds of other ways. People coming into science now have been raised in a communication culture. What they do in science will be determined not so much by their formal science education as by their experience as consumers of a very sophisticated communication system. This has to be taken very seriously.

Lastly, as scientists we must be very cautious in this new relationship with our public and we must not hang ourselves by making false claims of various kinds.

Index of Contributors

Subject Index